INNOVATION STRATEGIES AND ORGANIZATIONAL CULTURE IN TOURISM

This book analyzes the importance of innovation as the key driver of sustained success in the tourism industry and the knowledge sharing process. It explores its impact on innovation capability and innovation performance of organizations.

With 12 chapters written by 22 contributors, the book offers international reflections. The first part of the book focuses on innovation management strategies and analyzes the importance of innovation as the key driver of sustained success in the tourism industry. Tourism and hospitality firms develop innovations with specific objectives and goals in mind and have several approaches to measure performance accordingly. The second part of the book focuses on knowledge sharing through case studies. Chapters analyze the knowledge-sharing process and its impact on the innovation capability and innovation performance of the firms. The essence of knowledge sharing is to provide a framework for management in their attempt to develop and enhance their organizational capability to innovate in tourism.

This invaluable book provides both theoretical and practical insights into innovation strategies and knowledge sharing in tourism. It will interest students, scholars and researchers, as well as executives and practitioners in the field of tourism and hospitality management.

Marco Valeri received his PhD in Strategic Management and Organizational Behavior from "Tor Vergata" University, Italy. He is Associate Professor in Organizational Behavior at Faculty of Economics, Niccolò Cusano University, Italy. He is a Lecturer in Applied Organizational Behaviour, Xenophon College, UK. He is Visiting Professor at University of Information Science and Technology (UIST) "St. Paul The Apostle", Macedonia. He is Adjunct Professor at Faculty of Social Sciences and Leisure Management, School of Hospitality, Tourism and Events, Taylor's University, Malaysia. He is Associate Researcher in Strategy, Magellan Research Center, School of Management, Iaelyon Business School, Jean Monet University, France. He is Honorary Associate Professor, University of Pannonia, Hungary. His teaching and consultancy fields include strategic management, leadership development, cross-cultural management and international hospitality management. His research areas include sustainability and green practices, strategy implementation, knowledge management, family business, crisis management, information technology and network analysis. He has a long and extensive international academic experience and has taken visiting positions in several universities. He is a member of several Editorial Boards of international tourism journals, a reviewer and editor of several handbooks on entrepreneurship, tourism and hospitality management.

Routledge Insights in Tourism Series

Series Editor: Anukrati Sharma

Head & Associate Professor of the Department of Commerce and Management at the University of Kota, India

This series provides a forum for cutting edge insights into the latest developments in tourism research. It offers high quality monographs and edited collections that develop tourism analysis at both theoretical and empirical levels.

Diasporic Mobilities on Vacation
Tourism of European-Moroccans at Home
Lauren B Wagner

Overtourism and Cruise Tourism in Emerging Destinations on the Arabian Peninsula
Manuela Gutberlet

Pseudo-Authenticity and Tourism
Preservation, Miniaturization, and Replication
Jesse Owen Hearns-Branaman and Lihua Chen

Developing Industrial and Mining Heritage Sites
Lavrion Technology and Cultural Park, Greece
Taşkın Deniz Yıldız

Innovation Strategies and Organizational Culture in Tourism
Concepts and Case Studies on Knowledge Sharing
Edited by Marco Valeri

For more information about this series, please visit: www.routledge.com/Routledge-Insights-in-Tourism-Series/book-series/RITS

INNOVATION STRATEGIES AND ORGANIZATIONAL CULTURE IN TOURISM

Concepts and Case Studies on Knowledge Sharing

Edited by Marco Valeri

Routledge
Taylor & Francis Group
LONDON AND NEW YORK

Designed cover image: Henk Leerink

First published 2024
by Routledge
4 Park Square, Milton Park, Abingdon, Oxon OX14 4RN

and by Routledge
605 Third Avenue, New York, NY 10158

Routledge is an imprint of the Taylor & Francis Group, an informa business

British Library Cataloguing-in-Publication Data
A catalogue record for this book is available from the British Library

ISBN: 978-1-032-63332-9 (hbk)
ISBN: 978-1-032-63334-3 (pbk)
ISBN: 978-1-032-63337-4 (ebk)

DOI: 10.4324/9781032633374

Typeset in Times New Roman
by Apex CoVantage, LLC

CONTENTS

PREFACE

The twenty-first century tourism destinations have the imperative to innovate and remain competitive in an increasingly global competitive environment. Academics and managers have long been discussing on it. In tourism management literature, the majority of applications are concerned with individual organizations rather than the complex amalgams of organizations that characterize destinations.

Management literature in the tourism industry deals with the problems and methods of managing a tourist destination. The establishment of an authority of governance is of fundamental importance for the competitiveness of a tourist destination. The governance of the destination can be conceived as a system of strategic decisions and actions aimed at consolidating, and over time improving, the competitiveness of the destination, and defining clearly the organizational assets and shared development goals and knowledge among all parties in the tourism industry.

There are to date only a small number of examples and applications of knowledge sharing across destination networks. However, recognition of the significance of the approach is growing as practitioners recognize the value of knowledge sharing not just within the organization, but also through networks, and in particular the encouragement of partnerships within destinations. It is characterized by the fact that the early phases of knowledge management were characterized by the phrase "knowledge is power." The new thinking argues "sharing is power" and creates "communities of knowledge" at the destination level.

A concept that must be considered in understanding destinations as networks of organizations is that of the stakeholder. The concept is related to changing public sector governance as well as participatory management in the private sector. Stakeholders are the people who matter to a system. A stakeholder is any person, group or institution that has an interest in a development activity, project or program. This definition includes intended beneficiaries and intermediaries, winners and losers, and those involved or excluded from decision-making processes.

Data, information and knowledge flows in a destination network are relevant mechanisms for the general behavior of the system. Productivity, innovation and economic growth are strongly influenced by these processes, and the way in which the spread occurs can determine the speed by which individual actors perform and plan their future actions at the destination. The structure of the network will be influential in determining the efficiency of the destination's attempts to share knowledge and innovate.

The literature in this field has dealt with two main issues: the mechanisms and the processes of knowledge acquisition within a single stakeholder (e.g. a company, association or group) and the diffusion within the destination network formed by groups of stakeholders, based on their similarity (industrial clusters, for example), or because of their spatial location. The topology of the destination network formed by the different stakeholders and their formal and informal relationships has proved to be an important determinant when explaining the mechanisms by which ideas, information and knowledge "travel" from one element of the system to another.

The book *Innovation Strategies and Organizational Culture in Tourism. Concepts and Case Studies on Knowledge Sharing* is the result of reflections involving research studies of different nationalities. The book contains 12 chapters written by 22 authors located in seven different countries (Australia, China, Greece, Italy, Malaysia, Slovakia, Turkey) and affiliated with 15 different Universities. The book will be an opportunity to stimulate academic research, approaches and methods, developing a synthesized framework of innovation management process consisting of several categories: innovation strategies, knowledge sharing, organizational culture and structure, project management and commercialization in tourism.

The book is structured in two parts. The first part focuses on innovation management strategies. This part collects chapters that analyze the importance of innovation as the key driver of sustained success in the tourism industry. Tourist and hospitality firms develop innovations with specific objectives and goals in mind and have several approaches to measure performance accordingly. The second part focuses on knowledge sharing: case studies. This part collects chapters that analyze the knowledge-sharing process and its impact on the innovation capability and innovation performance of the firms. The essence of knowledge sharing is to provide a framework for management in their attempt to develop and enhance their organizational capability to innovate in tourism.

FIGURES

TABLES

CONTRIBUTORS

Ahmad Albattat is an Associate Professor in Graduate School of Management, Post Graduate Centre, Management and Science University, Shah Alam, Selangor, Malaysia. He is a Visiting Professor and External Examiner in Medan Academy of Tourism (Akpar Medan). He received PhD in Hospitality Management "Disaster and Emergency Planning and Preparedness" from University Sains Malaysia (USM), Malaysia. He worked as an Assistant Professor, Ammon Applied University College, Amman, Jordan. Senior Lecturer and Research Coordinator in School of Hospitality & Creative Arts, Management and Science University, Shah Alam, Selangor, Malaysia, and Researcher at Sustainable Tourism Research Cluster (STRC), Pulau Pinang, Malaysia. He was working for the Jordanian hospitality industry for 17 years. His teaching and consultancy fields include hospitality management, events management, research methodology, strategic management and final research projects. His research areas include hospitality management, hotel, tourism, events, emergency planning, disaster management, sustainable development goals and human resource. He is an active member of several Scientific and Editorial Board of International journals, reviewer, author and editor for book projects. His latest works have been published in the refereed international journals, conference proceedings, books and book chapters.

Eftychia Christina Aivaliotou, Directorate of Research, Dept. of Tourism Statistics, Ministry of Tourism of Greece. Hellenic Open University (HOU): Master on "Management & Business Administration". She is currently Deputy Head of the Department of Tourism Statistics in the Hellenic Ministry of Tourism. During the period 2013–2016, she had been appointed Director at the Office of the Greek National Tourism Organisation (GNTO) in Germany. She is a graduate of the Department of Economics of the University of Piraeus, of the National School of Public

Administration and Local Government and holds an MBA from the Hellenic Open University. She speaks fluently English, French and German.

Norhidayah Bt Azman is Senior Lecturer in Faculty of Business Management and Professional Studies, Management and Science University, Shah Alam, Selangor, Malaysia. She holds a doctoral degree in Business Management from Management and Science University (MSU). With a strong 7 years of industrial experience, her particular specialization is on digital marketing, big data, management, human resources, consumer behavior and marketing. She also received Certified Marketing Professional from American Industrial Certification Institute for completing marketing professional courses. She has participated and presented research papers at several International academic conferences. She is an active member of Scientific and Editorial Review Board for various SCOPUS journals.

Mahmut Baltacı received Phd in Tourism Management from "Mersin" University, Türkiye. He received his bachelor degree from "Mugla" University and master degree from "Selcuk" University. He is an Assistant Professor in Tourist Guiding Programme at Department of Travel, Tourism and Entertainment Services, at the Vocational School of Silifke, Tasucu of Selcuk University, Türkiye. His teaching and consultancy fields include travel management, ticketing management, tourist behavior, tourism marketing, gastronomy and tourist guiding. He has many book chapters, essays and conference papers about his teaching fields. He is also a member of Türkiye Tourist Guides Association as a tourist guide.

Stylianos Bouzis is a PhD candidate at the International Hellenic University in the field of Tourism and Hospitality Management and Certified Adult Educator (member of the Register of Instructors of Panteion University), Greece. He holds a bachelor's degree in Russian Language, Literature and Culture and a master's degree in Strategic Tourism Destination Management, specializing in Tourism Development Strategy. His research interests focus on the tourism development, special and alternative forms of tourism, glamping and tourism innovation.
He works as Assistant to the Development Director at Artion Conferences and Events, one of the most outstanding PCOs and DMSs in Greece.

Willem JL Coetzee holds a PhD in sustainable tourism. He has developed and taught several subjects in sustainable tourism and event management in South Africa (1996–2014), New Zealand (2015–2022) and now at Western Sydney University in Australia. His research focuses on (i) the resilience of the hotel industry in water-stressed destinations and the SDGs, (ii) attendee experiences at festivals and events and (iii) LGBTTQIA+ tourism and events. As an expert in the field, he supervised more than 20 Doctoral candidates to completion and is serving on several editorial boards, conference organizing committees, PhD workshops and academic writing retreats for early career researchers.

Jingjing Dai holds a PhD from South Korea and has worked in various cultural settings, such as China, Hong Kong SAR, the UK, South Korea, and the United Arab Emirates. She is an associate professor at the Office of Domestic and Foreign Affairs, Ningxia University, China. Currently, she works as the Chinese director at the Confucius Institute at the University of Dubai, UAE.

Bob Yi-Chen Duan is a lecturer at the School of Vocational and Technical Education, Yunnan Normal University, China. He received his education and practiced teaching on three continents, including Asia, Australia/Oceania, and Europe. The research area is cultural tourism. He is interested in social psychology's linkage to cultural phenomena, especially in the context of wine tourism.

Aslı Sultan Eren completed her undergraduate education at Sinop University, School of Tourism Management and Hotel Management, Department of Tourist Guidance and her master's degree at Nevşehir Hacı Bektaş Veli University, Institute of Tourism Research, Department of Tourism Guidance. As a PhD candidate, she is still continuing her education at Nevşehir Hacı Bektaş Veli University, Department of Tourism Guidance. Furthermore she working at Istanbul Esenyurt University as a lecturer. The author has an international book chapter, national and international papers and articles. Her research interests include tour guiding, tourist behavior, tourism ethics, gastronomy tourism and etc.

Gül Erkol Bayram is currently Associate Professor in School of Tourism and Hotel Management, Department of Tour Guiding, University of Sinop, Sinop, Turkey. Her doctorate is in Tourism Management from the Sakarya University, Turkey, and she completed her dissertation research on Tour Guiding in Turkey. Her core subjects are tourism, tour guiding, tourism policy and planning, and women studies. Erkol Bayram has also worked as a professional tour guide in the tourism sector. She has many book chapters and articles in the national and international arenas. She has been invited for many talks/lectures/panel discussions by different universities.

Tolga Gök is Associate Professor at the department of Hotel, Restaurant and Catering Services at the Vocational School of Silifke Tasucu, Selcuk University, Mersin, Türkiye. He serves as the Director of the Vocational School and the Head of the Hotel, Restaurant and Catering Services Department. With the approval of the Turkish Higher Education Institution, He worked at the Department of Travel Business and Tourism Guidance in Tourism Faculty of Kyrgyz-Turkish Manas University, Bishkek, Kyrgyzstan between 2016 and 2021. During this period, He served as vice dean for two years and dean for two years at the Tourism Faculty. He completed his Bachelor's degree (1998) and Master's degree (2002) in Tourism Management and his doctorate (2009) in Business Administration. His research interests are focused on tourism management, tourism behavior, gastronomy and cultural heritage. He has published articles, book chapters and conference papers about tourism.

Antonios Kritikos graduated from the Department of Social Anthropology of Panteion University of Social and Political Sciences in 2016. He holds a master's degree in business administration (MBA) from the University of West Attica in 2018. He also obtained a master's degree in "Innovation and Entrepreneurship in Tourism" (MSc) from the University of West Attica in 2019 with an oath rating of "excellent" in 2021. He is a PhD candidate of the National and Kapodistrian University of Athens, Department of Business and Organizational Management, Greece. In 2019, his award at the pan-Hellenic university level by the Hellenic Association of Professional Congress Organizers is a benchmark based on the "Waste Management and the production of electricity from waste as a solution for a sustainable tourism development." In 2023, he was awarded by the sustainable development journal *SDCT-Journal* as the best research study of the 28 issues of the journal with a case study of the Acropolis Museum entitled "Digital technology as a means of promoting cultural tourism in Greece."

Jeetesh Kumar is a Senior Lecturer at the School of Hospitality, Tourism & Events; Hub Leader at the Responsible Tourism for Inclusive Economic Growth" hub under the Sustainable Tourism Impact Lab and Associate Director for the Centre for Research and Innovation in Tourism (CRiT) at Faculty of Social Sciences and Leisure Management, Taylor's University, Malaysia. His research interests include sustainable tourism, community participation, economic impacts, and economic modelling.

Emily Ma is a professor at the School of Hospitality and Tourism Management at the University of Surrey, UK. Her research area includes organizational behavior, customer experience management and women in leadership.

Ladislav Mura is Associate Professor at the Department of Tourism, Faculty of Commerce, University of Economics, Bratislava, Slovakia. He is Slovak economist and expert on small and medium enterprises, family businesses, international business management and the phenomenon of internationalization of business. He is the author of three domestic scientific monographs, two foreign scientific monographs and some papers in scientific journals (with impact factor, too), including foreign journals and contributions to foreign and domestic conferences.

Panagiota Dionysopoulou, Director General of Tourism Policy, Ministry of Tourism of Greece, Adjunct Professor School of Social Science, Post-graduate courses "Tourism Business Administration Msc", Hellenic Open University (HOU). He is the Director General of Tourism Policy at the Ministry of Tourism of Greece. She holds a PhD in Tourism Economics and Politics. She has postgraduate studies in Public Economics and Politics and International and European Relations. She is a graduate of Athens University of Economics and Business.

She has many years of teaching experience in Higher Education. In addition to her teaching experience, she has held management positions, as Director and General Director at the Ministry of Education.

She has participated in the elaboration of tourism development studies and published original papers in scientific conferences and journals. She is the author of the book "European Tourism Policy" and has participated in collective books.

Panoraia Poulaki is Academic Laboratory Teaching Staff in the Department of Economics and Management of Tourism of the University of the Aegean, Greece. She also teaches in the MSc Program in Strategic Management of Tourism Destinations and Hospitality Enterprises of the University of the Aegean. She holds a first degree in History and Archaeology from the National and Kapodistrian University of Athens, Master of Science (MSc) in Planning, Management and Policy of Tourism from the University of the Aegean and a Doctor of Philosophy (PhD) in Tourism from the same University. She has written a book about the contribution of the Ottoman Monuments of Chios to the development of cultural tourism, she has published research papers in scientific journals, and she has presented her research work in international conferences. Her research interests focus on the development, planning and management of tourism, with emphasis on special and alternative forms of tourism. In 2023, she was awarded by the sustainable development journal *SDCT-Journal* as the best research study of the 28 issues of the journal with a case study of the Acropolis Museum entitled "Digital Technology as a Means of Promoting Cultural Tourism in Greece."

Theodoros Rachiotis is an educator in the private sector and a PhD candidate at the School of Physical Education and Sports of the Kapodistrian University of Athens, with a subject related to both sports psychology and sports research methodology, Greece. He holds a first degree in History-Archaeology from the Aristotle University of Thessaloniki and a Master of Science (MSc) in Planning, Management and Policy of Tourism from the University of the Aegean. He has published research papers in scientific journals and has presented his research work in national and international conferences. His research interests focus on the development, planning and management of tourism, but also on sports psychology.

Anastasios E. Thomaidis is Athens Lawyer appointed to the Supreme Court of Greece. He holds PhD in Civil and Tourism Law (Law School, Aristotle University of Thessaloniki), Adjunct Lecturer in Civil Law in the Law School of the European University of Cyprus and Legal Advisor to the Ministers of Tourism (2019–2021) and Citizen Protection (2021–2023). He is Head of the Energy Sector of the Greek governing political party of New Democracy.

Marco Valeri is Associate Professor in Organizational Behavior at Faculty of Economics, Niccolò Cusano University, Italy. He is a Lecture in Applied Organizational

Behaviour, Xenophon College, UK. He is Visiting Professor at University of Information Science and Technology (UIST) "St. Paul The Apostle", Macedonia. He is Adjunct Professor at Faculty of Social Sciences and Leisure Management, School of Hospitality, Tourism and Events, Taylor's University, Malaysia. He is Associate Researcher in Strategy, Magellan Research Center, School of Management, Iaelyon Business School, Jean Monet University, France. He is Honorary Associate Professor, University of Pannonia Hungary. His teaching and consultancy fields include strategic management, leadership development, cross-cultural management, and international hospitality management. His research areas include sustainability, strategy implementation, knowledge management, family business, crisis management, information technology and network analysis. He is a member of several Editorial Boards of international tourism journals, reviewer and editor of several handbooks on entrepreneurship, tourism and hospitality management.

Nikolaos Vasilakis is a teacher in secondary education in Chios of Greece and a researcher in history, culture and Greek language. He holds a degree in History from the Ionian University (Corfu) of Greece. He has published research papers in scientific journals and has presented his research work in international conferences. His research interests focus on history, culture and Greek language. In 2023, he was awarded by the sustainable development journal *SDCT-Journal* as the best research study of the 28 issues of the journal with a case study of the Acropolis Museum entitled "Digital Technology as a Means of Promoting Cultural Tourism in Greece."

Tibor Zsigmond is an assistant professor at the Department of Management, Faculty of Economics and Informatics, János Selye University, in Komárno, Slovakia. He completed his doctoral studies at this faculty too, defending his dissertation in August 2021. He published more than 80 papers, some of which are listed in international databases (WoS, Scopus). He has also participated in various domestic and foreign conferences in recent years, including several times in Bratislava, Subotica, Hradec Králové, Prague, Pula and Portorož. He was the winner of "Danubius Young Scientist Award" in 2022.

PART 1

Innovation management strategies

1

ROLE OF NEW ERA ON INFORMATION MANAGEMENT OF TOURIST GUIDES

Aslı Sultan Eren, Jeetesh Kumar, Gül Erkol Bayram and Marco Valeri

Introduction

Tourist guiding profession, which is one of the sub-professional branches of tourism, which is a service sector, is also among the professions that benefit from technological developments. At first glance, there is a belief that the impact of technology will be minimal in the tourist guiding profession, where human relations are at the forefront. However, in addition to the various applications that can be downloaded to mobile communication devices and used by people in the destinations they visit, the increase in the number of robots providing guidance services in closed areas such as museums and exhibitions in various countries of the world since 1991 shows the opposite of this belief. A guide robot that provides guidance in any museum or exhibition hall visited by domestic and foreign tourists in Turkey has not yet been developed. However, in addition to the "audio guide" application called an audio guide, which has been started to be used in order to provide more comprehensive information in different languages to the tourists visiting the museums and ruins, as well as the mobile microphone and earphone assembly known as the "head-set" working with radio frequency to be used in group tours in Turkey. It also exemplifies the effects of technology on the tourist guiding profession (Yıldız, 2019).

Headset and audio guide systems used in Turkey facilitate the work of tourist guides. However, robot tourist guides used in various countries are used as a substitute for human tourist guides in museums or exhibition halls where they provide guidance. In other words, there are robots as well as humans in the tourist guide profession in these countries. This puts human tourist guides at risk of losing their jobs. Unfortunately, there is almost no study on the current and future roles of the developments in the field of technology in the domestic and foreign literature,

DOI: 10.4324/9781032633374-2

especially the robot tourist guides in the tourist guiding profession. Existing studies generally focus on the technical features of robot guides. This study aims to raise awareness about the existence of technology-based guidance activities, the importance of which is increasing day by day in the tourist guiding profession.

General review of tour guides: definitions, challenges, current issues

It is seen that tourist guides in the world come together with many different institutions and organizations and try to act jointly. Firstly, the WFTGA definition is "*A person who guides visitors in the language of their choice and interprets the cultural and natural heritage of an area, normally with a site-specific qualification usually awarded and/or recognised by the appropriate authority*" (WFTGA, 2023). Then, when the definition of FEG (European Federation of Tourist Guide Associations) is examined, the tourist guiding profession has been defined as "*Historically Tourist Guiding has developed for decades as a distinct profession within the tourism industry alongside other complimentary professions that also contributes to the visitors' overall experience*" (FEG, 2022). After examining the definition of these globally accepted organizations, the definitions of tourist guides in the literature will be examined.

Tour guide profession defines it as the person who accompanies the tourists from the moment they arrive until the moment they leave the country, gives the information that should be given according to the nature of the tour, takes them to places where accommodation, transportation, food and entertainment services will be purchased, and ensures that the tourist can benefit from touristic services (İlhan et al., 2023). Tourist guiding profession is a profession that requires individuals, who want to visit a destination, get information about it and have a different experience, provide information about the natural, historical and cultural heritages in the region, keep their information always up to date, lead the tourists and act as an ambassador between the tourist and the destination (Cohen, 1985; Eren et al., 2020).

Tourism service is an important sector that offers people the opportunity to spend their free time with quality (Sak et al., 2022). The duties and responsibilities of tourist guides and their characteristics are also of great importance in terms of tourism because tourist guides are the person who will establish a link between the tourist and the destination (Ap & Wong, 2001). For this reason, the success and failure of the relationship between the tourist and the destination are associated with the tourist guide (Zhang & Chow, 2004).

Tourist guide is a profession that actually carries its duties in its name because of the word "guide" in its name. When a mentor is mentioned, the definition of "guiding person" comes to mind first, followed by the definition of "mentor" (Cohen, 1985). In fact, these two definitions are among the primary duties of tourist guides. Tourist guides have many responsibilities before and after the tour. One of them is the responsibility for the value of the time tourists spend at the destination (Syakier &

Hanafiah, 2021). Thanks to the knowledge and strong communication skills of the tourist guides, he is responsible for solving the problem and maintaining the peace of the tour in case of an unexpected crisis situation during the tour (Christie & Mason, 2008). However, it is not possible to limit the duties of a tourist guide to this extent. Tourist guides are in contact with many people and institutions such as tourism stakeholders, local people, tourists and local governments.

In Figure 1.1, Cohen (1985) gathered the duties of tourist guides under four main headings in his study. He allocates the multiplicity of the roles tourist guides play into two areas he calls the areas of leadership and mediatory. These two areas have both an outward and an inward aspect, in other words they have duties outside the tour and duties within the group, these constitute the four main components of the duties of tourist guides: its instrumental role, social role, interactional role and communicative role (Ap & Wong, 2001; Cohen, 1985; Syakier & Hanafiah, 2021; Weiler & Black, 2015; Zhang & Chow, 2004). His instrumental role in the field of outward and leadership is related to the problem of the tour guide that he has gone on without any problems. For this reason, it is seen as one of the "primary" roles. Instrumental role; It consists of direction, access and control components. Its outward and interactional role from the mediation field; It draws attention to the role of mediator between tourists and local people, tourism businesses and official institutions in the destination. The interactive role includes the components of representing and organizing. The social roles of the tourist guide, which is seen as an inward role from the leadership area, refer to the strong communication between the tour participants and the ability to keep the energy of the tour high and to solve the problems in the tour professionally. Stress or crisis management, in-group harmony, high energy and recreational activities are the components that make up the social role of the tourist guide. Finally, there is the communicative role, which is inward-oriented from the mediation field. The communicative role represents the skills they use during their communication with tourists and includes the ability to select, inform, interpret and narrate (Ap & Wong, 2001; Cohen, 1985).

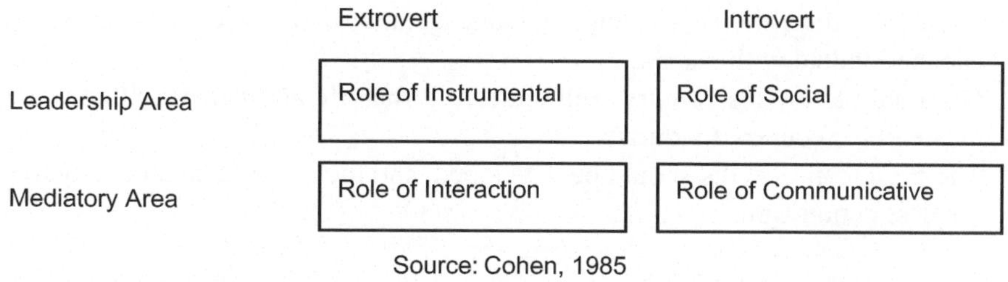

FIGURE 1.1 Schematic representation of the principal components of the tourist guide's role

Tourist guiding is a very valuable profession. Although tourist guides have important duties, they should also have the features to cope with the tasks. Tourist guides should act as a bridge between local people–tourist and tourist–destination, sometimes by acting as a mentor or a storyteller, sometimes as a translator, conveying the knowledge they have like a teacher. For this reason, he should have the characteristics of a cultural ambassador. In order to keep the excitement and energy in the tour high, there should be an entertainment actor at times, and sometimes he should guide tourists, guide them, and take the right decisions on their behalf, as the real meaning of the word "guide" in his name (Chang, 2014; Chang et al., 2020; Cohen, 1985; Eren et al., 2020; Huang et al., 2010; Pearce, 1984; Weiler & Black, 2015). Therefore, it is very important for tourists to be able to convey the characteristics of tourist guides and the situations experienced during the tour, to show themselves and to be approved by tourists (Pearce, 1984; Pearce & Lee, 2005). Çimrin (1995) listed the features that should be in a tourist guide as follows:

- He should take care of his clothes
- Speak fluently and clearly in the language he/she serves
- Must be a polite, fun and serious individual with boundaries
- Avoid arguing with the tourist
- He should not turn his back on the tourists while making his presentation
- Must have etiquette rules
- Avoid unnecessary intimacy with tourists, not to talk too much and bore the tourist
- Should not behave like "I know everything"
- Should deal with all the problems of tourists
- Must be honest
- Must be on time
- Should not make changes in the tour program according to their personal interests
- Keep your cool in times of crisis
- Must be sociable, not shy
- Treat each of the tourists equally
- Should not discriminate on religion, language, race and gender and should treat each individual with respect
- He should know his country well and avoid rhetoric and behavior that would discredit his country to tourists
- The presentation skills should be developed, and they should be able to market the destination well

In this way, it is possible to list the features that a tourist guide should have. To enable the tour guide to receive positive feedback from the tourists if he or she successfully transfers these features to the tour (Çimrin, 1995; Syakier & Hanafiah, 2021).

Tour guiding & information management with perspective of SWOT Analysis

Information management has been defined by the American Center for Production and Quality as systematic approaches to ensure that knowledge is disseminated at the right time and to the right person so that it can emerge and create value (Buckman, 2004, p. 17). Information management is basically a discipline that updates the ever-increasing information capacity in the organizational environment, makes the information available, defines the necessary processes to reach the necessary information, and ensures that the necessary information is shared with the company employees (Harrison & Kessels, 2004, p. 39).

According to Awad and Ghaziri (2004), knowledge management is an interdisciplinary model that carries knowledge within the framework and focus of attention of the organization. It is based on many disciplines such as economics, psychology and information management. It is the ultimate advantage in today's organizations. Parts of it include people, technology and processes. It has a positive effect on business processes. The purpose of knowledge management is to capture what is needed in a process and to support knowledge workers in sharing this knowledge. If an organization can capture and disseminate knowledge, the benefits will be endless (Daghfous, 2003).

When the information management is examined in terms of tourist guides, it has turned to different channels in line with the developing technology and diversifying needs. In tour guiding, at the stages of knowledge management, the most appropriate way to get the point across is to set high-level principles rather than elaborate tactics. Once a tour of basic principles regarding knowledge management have been agreed upon, detailed approaches and plans can be created based on these principles. These principles are as follows (Barutçugil, 2002, p. 85):

- Information management in tour guide is a continuous process
- Information management in tour guide is expensive to use
- Effective information management in tour guide requires joint solutions of people and technology
- Confidence and belief are required for the success of information management in tour guide
- It is very important to follow the latest technological developments and a comprehensive team for information management applications in tour guidance
- Information management in tour guide means the improvement of knowledge of business processes
- Sharing and using information through technology are often unnatural actions. It is necessary to adapt the profession of tour guide and keep up with the times

- Information management in tour guide makes use of knowledge maps
- Information management in tour guide is a long process. It is not enough for the tourist to just be aware of the information and reach the information. It is necessary to get the right information from an expert in the field at the right time

As a result of the SWOT analysis, the strengths, weaknesses, advantages and threats of technological developments to the tourist guiding profession were determined. These are;

Strengths

- Tour guides have the opportunity to go on more tours due to the increase in people's participation in the tour thanks to the developments in transportation technology
- Thanks to transportation technologies, travels become safer for both tourists and guides
- Thanks to communication technologies, the tourist–guide relationship becomes stronger before and after the tour than in the past
- Minimizing the problems experienced in finding directions and alternative routes thanks to navigation technology
- Elimination of the need for tourist guides to carry auxiliary resources with them

Weaknesses

- Advanced information technologies bring information pollution
- The possibility of errors in the use of technology causing serious problems
- The possibility of guides or tourists not being able to fully master the technology
- Rapid developments and innovations in technology constitute a significant expense

Opportunities

- Compared to the past, tourist guides can become more knowledgeable culturally and geographically in a shorter time
- Thanks to advanced communication networks and social media, guides have the chance to get extra tours by contacting customers directly
- Technology acting as a driving force for guides to prepare themselves in a shorter time about geographies they have not been to before

Threats

- The risk that technology-based guidance practices may replace guides in future

- The possibility that guides who cannot adapt to the developing technology will fail in competition
- The probability of the tour being interrupted as a result of not obtaining the expected benefit from each technological product

Tour guides in new area: technology as promoter and helper in guided tours

The use of the Internet has also become widespread due to the widespread use of computers. Today, thanks to mobile communication devices such as internet-compatible smartphones or tablets that almost everyone has, people have the opportunity to access information whenever they want without the constraints of time and space (Demir & Demir, 2022). These devices, thanks to the fact that they have a navigation feature with various GPS-supported applications downloaded from the internet, have made a significant contribution to the fact that people can easily reach the place they want to go in a city or country they are unfamiliar with. Although these developments do not have a direct impact on the tourist guiding profession, they have reduced the dependence of tourists visiting a country on interpreter tourist guides or guidebooks (Rui et al., 2023). Science and technology have made their presence felt in the tourist guiding profession with robot tourist guides as well as various applications downloaded to mobile devices. In time, there have been developments in the features of robots as well as their usage areas. Developments in the field of artificial intelligence continue to contribute significantly to taking important steps toward giving robots human characteristics. Thanks to artificial intelligence studies, there is talk of the existence of humanoid robots, which are called "humanoids" and have features close to those of a human (Vadreas et al., 2022; Rui et al., 2023).

Guidance profession also takes its share from technology and innovation. Advanced GPS-assisted map applications, applications that provide detailed, written and audio explanations about museums and sites can now offer tourists an alternative to travel without a guide, even in the destinations they visit for the first time (İstanbullu Dinçer & Çakmak, 2015). In addition, many different technological developments offer important outputs for the guide. When the literature on the interaction between tourism guidance and technological developments is examined, no direct publication on tourist guidance and technological developments has been found. In the studies (Kenteris et al., 2009; Vansteenwegen & Van Oudheusden, 2007; Souffriau et al., 2008), tourist guides use GPS, internet, etc., during the tour. There are research works on the structure of technologies, there is no evaluation of these technologies in terms of tourism, and studies in the field of engineering are carried out on the algorithms of existing technologies. However, today, as in all sub-sectors of the tourism industry, it is striking that tourist guides also make extensive use of technology. The "head set" system, GPS-navigation technologies, smartphones, and developments in transportation systems, which have been put into practice in museums and ruins, affect the guides in performing their profession; on the one hand, they also affect the tourists participating in the tour very positively, and it is thought to increase the participation or efficiency of the tour.

The position of the use of technology in the profession of tour guide today is such that it allows one to visit important tourist points that are known and expected to be visited without a tour guide. These applications are proof that the developing technology and information management systems have been directly transferred to the tour guide profession, but in today's world, there is more development in the field of technology every year. This situation creates the need for continuous renewal of the technology integrated into the tour guide profession. At the same time, the changing perception of tourism after the COVID-19 epidemic (Eren et al., 2020) and the increase in participation in special interest tourism necessitated the continuous spread of technological developments integrated into tour guiding activities to new places. With these technological developments, tour guides will only narrate to fewer groups who want tour guided tours, and in a way, the workforce of the tour guide will also decrease (İstanbullu Dinçer & Çakmak, 2015).

Today, with the use of humanoid robots that can provide guidance both indoors and outdoors, instead of robots that provide guidance only in indoor places such as museums and exhibition halls, the human–machine relationship has been added to the tourist guide profession, where the human–human relationship is at the forefront, and accordingly, the tourist guiding profession has become more important (Kenteris et al., 2009; Vansteenwegen & Van Oudheusden, 2007; Souffriau et al., 2008).

Conclusion

It is an indisputable fact that the developments in the field of science and technology have an impact on many professions within the scope of the service sector as well as the production-oriented sectors today, as in the past. While the effect of technology is expected to be at a certain level in the service sector, where human relations are at the forefront compared to the manufacturing sector, the developments in advanced technologies such as robots, software and artificial intelligence have led to an increase in the effect of technology on the professions where human relations are at the forefront. As a result of these developments, concepts such as digital nurse, robot teacher, robot pharmacist, robot receptionist, and robot tourist guide are more frequently encountered on various platforms. In other words, the human–machine relationship has been added to the human–human relationship, which are the basic elements of the service sector.

Now, tourists have the same expectations in the guidance service as if they expect qualified and special service in the hotels they stay or in the travels they make before going on cultural tours. Groups, who acquire basic information about the destination they will go through the Internet with their own efforts, ask for more specific information from the guides, forcing the guide to be more equipped. At this point, just as the tourists use technology to be more knowledgeable and conscious, the guides cannot also improve themselves more easily, go on tours more prepared, and make more comebacks if they are not integrated into the expectations with technology. In addition, technologies are rapidly advancing toward becoming an alternative to guidebooks. Now, tourists can make their own tour plans by downloading the necessary free applications to their smartphones or portable tablet computers, and they can also choose the local tastes to eat or the places to shop without the support of the guides, and create their own tour plans through these applications. From this point of view, technology is beginning to restrict the work area of guides on the one hand, and it can also provide unguided tours to individuals who like to travel without being tied to any group or guide. Within the scope of all these rapid developments, information society and the reality of the globalizing world, the importance of technology use by the guides emerges as an issue that should be realized at the level of the union.

References

Ap, J., & Wong, K. K. F. (2001). Case study on tour guiding: Professionalism, issues and problems. *Tourism Management*, *22*(5), 551–563.

Awad, E., & Ghaziri, H. (2004). *Knowledge management*. Prentice Hall Publishing.

Barutçugil, İ. (2002). *Bilgi Yönetimi*. Kariyer Yayıncılık.

Buckman, R. (2004). *Building a knowledge – Driven organization*. McGraw-Hill Companies Pub.

Chang, K. C. (2014). Examining the effect of tour guide performance, tourist trust, tourist satisfaction, and flow experience on tourists' shopping behavior. *Asia Pacific Journal of Tourism Research*, *19*(2), 219–247.

Chang, S. Y., Tsaur, S. H., Yen, C. H., & Lai, H. R. (2020). Tour member fit and tour member–leader fit on group package tours: Influences on tourists' positive emotions, rapport, and satisfaction. *Journal of Hospitality and Tourism Management*, *42*, 235–243.

Christie, M. F., & Mason, P. (2008). The good guide: Identifying and engendering generic skills in the training of tourist guides. In *Tourism development revisited: Concepts, issues and paradigms* (pp. 341–350). Sage Publication Ltd.

Çimrin, H. (1995). *Tourism and ABC of tour guiding*. Akdeniz Publish.

Cohen, E. (1985). The tourist guide: The origins, structure and dynamics of a role. *Annals of Tourism Research*, *12*(1), 5–29.

Daghfous, A. (2003). How to make knowledge management a firm's core capability. *Journal of Knowledge Management Practice*, 4.

Demir, Ş., & Demir, M. (2022). Evaluation of digital tour guide technology and applications by professional tourist guides. *Scientific Collection "InterConf"*, *135*, 51–61.

Eren, A. S., Sak, M., & Erkol Bayram, G. (2020). The changing dynamics of the tourist guidance profession after Covid-19. *Cape Comorin Journal*, *3*(2), 81–87.

FEG. (2022). *Tourist guiding as a unique and distinct profession*. www.feg-touristguides.com/about.php (Accessed 29 July 2022).

Ham, S. H., & Weiler, B. (2002). Toward a theory of quality in cruise-based interpretive guiding. *Journal of Interpretation Research*, *7*(2), 29–49.

Harrison, R., & Kessels, J. (2004). *Human resource development in a knowledge economy*. Palgrave Macmillan Pub.

Huang, S., Hsu, C. H. C., & Chan, A. (2010). Tour guide performance and tourist satisfaction: A study of the package tours in Shanghai. *Journal of Hospitality and Tourism Research*, *34*(1), 3–33.

Ilhan, I., Ayyildiz, S. Ç., & Eren, A. S. (2023). Bibliometric analysis of national articles and theses on tourist guiding professional ethics. *Journal of Tour Guiding (JOTOG)*, *6*(1), 1–11.

Istanbullu Dinçer, F., & Çakmak, T. F. (2015). *Evaluation of surveillance society management in terms of tourism*, I. Eurasian International Tourism Congress 28–30 May 2015 Konya (Vol. 2, pp. 178–188).

Kenteris, M., Gavalas, D., & Economou, D. (2009). An innovative mobile electronic tourist guide application. *Personal and Ubiquitous Computing*, *13*, 103–118.

Pearce, P. L. (1984). Tourist-guide interaction. *Annals of Tourism Research*, *11*, 129–146.

Pearce, P. L., & Lee, U. I. (2005). Developing the travel career approach to tourist motivation. *Journal of Travel Research*, *43*(3), 226–237.

Rui, M., Zhixiang, W., Yu, L., Wenkai, P., & Meilin, L. (2023). Design and implementation of tour guide robot for red education base. *Journal of Artificial Intelligence Practice*, *6*(3), 76–80.

Sak, M., Eren. A. S., & Erkol Bayram, G. (2022). Role of event tourism in economic development. In S. Arora & A. Sharma (Eds.), *Event tourism in Asian countries challenges and prospects içinde*. Apple Academic Press.

Souffriau, W., Vansteenwegen, P., Vertommen, J., Berghe, G., & Oudheusden, D. (2008). A personalized tourist trip design algorithm for mobile tourist guides. *Applied Artificial Intelligence*, *22*, 964–985.

Syakier, W. A., & Hanafiah, M. H. (2021). Tour guide performances, tourist satisfaction and behavioural ıntentions: A study on tours ın Kuala Lumpur city centre. *Journal of Quality Assurance in Hospitality and Tourism*, *23*(3), 597–614.

Vadreas, A. K., Purwasih, R., Saputra, H., & Stephane, I. (2022). Design and build an eco edu tourism application using a virtual tour guide in the thematic eco Enzyme village, Padang City. *Indonesian Journal of Community Services Cel*, *1*(2), 108–116.

Vansteenwegen, P., & Van Oudheusden, D. (2007). The mobile tourist guide: an OR opportunity. *OR Insight*, *20*, 21–27.

Weiler, B., & Black, R. (2015). The changing face of the tour guide: One-way communicator to choreographer to co-creator of the tourist experience. *Tourism Recreation Research*, *40*(3), 364–378.

WFTGA. (2022). *What is a tourist guide?* https://wftga.org/about-us/what-is-a-tourist-guide/ (Accessed 29 July 2022).

Yıldız, S. (2019). The rise of robot guides in tourist guidance profession. *Suleyman Demirel University Visionary Journal*, *10*(23), 164–177.

Zhang, H. Q., & Chow, I. (2004). Application of importance-performance model in tour guides' performance: Evidence from mainland Chinese outbound visitors in Hong Kong. *Tourism Management*, *25*, 81–91.

2

IMPACT OF EMOTIONAL INTELLIGENCE ON KNOWLEDGE SHARING AMONG EMPLOYEES WORKING AT SMEs IN THE TOURISM AND SERVICE SECTOR

Ladislav Mura and Tibor Zsigmond

Introduction

In recent decades, companies recognized that knowledge is crucial to maintain a favorable position of the company on the market (Milczarek et al., 2015; Szabó-Szentgróti et al., 2019; Zakrzewska et al., 2022; Sarihasan et al., 2022). However, this is not enough for success, as other factors also influence the success of organizations. Specific knowledge is a base to get a competitive advantage. In the case of SMEs, knowledge plays a significant role, and the primary source of knowledge are the employees themselves (Ključnikov et al., 2022; Ključnikov et al., 2021; Civelek et al., 2021; Wach et al., 2023). It is not enough to acquire the knowledge that is important for organizations; it is necessary to ensure that it is available to employees in the organization who will utilize it.

Furthermore, it is essential to use the knowledge already acquired again in future. These processes require a certain level of competence, which is why knowledge organizations were created and are involved in the activities mentioned earlier. Companies that do not manage knowledge properly can face great difficulties. Knowledge "cannot manage itself," so it may happen that due to careless knowledge management, it remains hidden within the company for people who need it. In such cases, the necessary knowledge must be acquired again. This means costs (money, time and energy) for organizations. Being a knowledge organization pays off in the long run. However, it is crucial for companies to hire individuals who are willing to share their knowledge. We believe that emotional intelligence can be a factor influencing the flow of knowledge. Furthermore, emotional intelligence also plays a significant role in building trust within the organization (Strugar Jelača et al., 2022).

DOI: 10.4324/9781032633374-3

These thoughts inspired us to address this issue. Our primary goal is to examine the impact of emotional intelligence on knowledge sharing within the organization. In order to formulate an objective opinion, we also conducted qualitative and quantitative research, the results of which are presented in the empirical part of the study. Based on these results we formulated recommendations that can be applied in practice and can have a positive impact on knowledge sharing in the company.

In addition, it is essential to emphasize the situation of SMEs and companies operating in the tourism industry and hospitality during the pandemic (Perovic et al., 2021; Kumar et al., 2021; Ajide, 2022; Bozsik et al., 2023; He et al., 2022). The tourism sector and the related services are closely tied to physical contact with people (Štefko et al., 2020; Valeri & Baggio, 2021; Yin et al., 2020; Tovmasyan, 2023). That is why emotional intelligence also plays a vital role in everyday social interactions (Santos et al., 2022). The closures and social distancing as a reaction to COVID-19 made the situation of these companies extremely difficult. Some companies "survived" generating losses (e.g. financial, knowledge-related), while others went bankrupt. After experiencing the difficult period of the pandemic, the sector faced an even more challenging period – the Russia–Ukraine war faded the optimism of people. The increasing energy and transport costs combined with increasing living costs became a burden for both the employer and the employees.

Theoretical background

Knowledge and knowledge organization

In order to understand knowledge organizations, clarifying the concept of knowledge is vital with regard to knowledge management. No uniformly accepted definition describes the exact, unambiguous meaning of knowledge. Several researchers have already interpreted the given term based on different approaches, and different definitions have been formulated. The work of Penrose (1959) can be highlighted first. In the book "The Theory of the Growth of the Firm," the author examines the circumstances and factors that influence the growth of companies. The process presented in his work resulted in various debates, as well as formulation of new theories and further research. The modern resource-based company theory was developed by Wernerfelt (1984). It is closely related to the works of Rumelt (1984), but the role of Penrose, who formulated the basis of this theory cannot be questioned either. Based on Penrose (1959, p. 24): "A company is a collection of productive resources, in which administrative decisions determine the distribution among different areas of use" (Kapás, 1999; Michie, 2019).

According to Penrose (1959), some of these resources are physical, while human capital is other form of resource. Penrose also mentioned the growing importance of knowledge. Knowledge is only a "foggy" concept that cannot be perceived as a solid component (Kapás, 1999; Michie, 2019).

Foss (1997) and Hodgson (1998a, 1998b) believe that Adam Smith is a forerunner of the resource-based theory, and based on his theory they published their work "Wealth of Nations" in 1776. According to Kapás (1999); Michie (2019), employees can access knowledge through learning and accumulate knowledge to develop their abilities.

At the end of the twentieth century, additional studies and theories related to the topic were elaborated. These include the work of Davenport and Prusak (1998). The authors define knowledge as a set of experiences, values and information. Aydin (2022) examines the topic, too.

According to Li et al. (2008) and Wierzbiński et al. (2023), knowledge is made up of experience, perception, insight and foresight. In other words, knowledge is a unique blend of intuition and intellect that enables individuals to understand how they should act.

It is important to mention Drucker's (2011) definition, which says: "Traditional factors of production – land (natural factors), labour and capital – have not disappeared. They have become secondary. Knowledge gained a priority importance and became a tool to achieve social and economic results" (Drucker, 2011, p. 38). Based on this, capital or natural resources are not the primary economic resource. Knowledge as an asset has replaced them (Zamir, 2019).

Emotion and emotional intelligence

The word "emotion" comes from the Latin word "motore" and the prefix "e"; thus, it means "to move." Therefore, the meaning of the word can be interpreted as emotions that make us do something. According to Neale et al. (2009), emotions influence our decisions, judgments and actions. According to Vágnerová (2005), emotions make it easier to adapt to certain situations and life circumstances.

Emotional intelligence as a concept dates back to the 1990s; however, specialists were concerned with the topic even before that. Thorndike (1920) even spoke of "social intelligence." He believed that emotional intelligence is an ability that can help individuals perceive and understand the behavior of other individuals. Emotional intelligence makes people more successful in social relationships. The creator of the WAIS, Wechsler (1955), emphasized the so-called "non-intellectual factors." These covered the social skills of individuals.

Later on, professionals in the field of psychology carried out numerous types of research. Among them, Beldoch (1964) and Leuner (1966) were the first to use the term "emotional intelligence." Payne (1986) also addressed the topic in his doctoral dissertation. Payne drew attention to the importance of emotional intelligence and its development. According to him, the development of emotional intelligence should begin in childhood. The only problem with the work of Beldoch, Leuner and Payne is that they did not deal enough with the concept of emotional intelligence itself, so none of them formulated a precise definition, nor did they develop a unified theory related to the concept.

It is also important to mention Gardner's (1983) theory of multiple intelligences, which has been presented earlier. Among the eight types of intelligence he described, we must emphasize personal (intrapersonal) and social (interpersonal) intelligence. These two types of intelligence help us understand our own and other people's feelings, and thus, they help us manage social situations better. The theory of Lane and Schwartz (1987) was published in the 1980s, which introduced the concept of "emotional awareness." They set up a model in which emotions are differentiated according to their complexity and presented all this as a development theory. Bar-On (1997) introduced the "emotional quotient" concept. According to him, emotional intelligence is "a set of non-cognitive abilities, competencies and skills that influence the individual to cope with the demands and pressure of the environment" (Bar-On, 1997, p. 14).

A significant breakthrough was introduced in 1990, as Salovey and Mayer (1990) defined the concept of emotional intelligence. They focused on Gardner's concepts of intrapersonal and interpersonal intelligence. Based on their definition, emotional intelligence is a skill that enables individuals to observe themselves. The model created by these two authors will be described in detail later. Saarni (1990) presented his theory the same year, and called his concept "emotional competence theory." Based on this, there are eight social-emotional competencies, which are interrelated. These are as follows:

> 1) awareness of one's own emotions, 2) ability to discern and understand other's emotions, 3) ability to use the vocabulary of emotion and expression, 4) capacity for empathic involvement, 5) ability to differentiate subjective emotional experience from external emotion expression, 6) adaptive coping with aversive emotions and distressing circumstances, 7) awareness of emotional communication in social relationships, and 8) capacity for emotional self-efficacy.
>
> *Kun (2011, p. 36)*

It is also important to mention the work of Goleman (1995), which popularized the concept of emotional intelligence. According to him, emotional intelligence is a set of abilities that include self-discipline, enthusiasm, persistence and the ability to motivate ourselves. Thanks to his work, the term of emotional intelligence became widespread. Due to his success, many have mistakenly credited Goleman with introducing the concept of emotional intelligence. In 1995, the Time magazine drew attention to the importance of emotional intelligence and said that it is the best predictor of success in people's lives, which means that EQ can be more important than IQ (Mayer, 2015; Garamvölgyi & Rudnák, 2015).

Sternberg (2000) defines emotional intelligence as an ability to recognize and express feelings, to be able to incorporate feelings into thoughts, understand feelings, and regulate the feelings of the individual.

There is still no agreement regarding the concept. Brackett and Mayer (2003) pointed out that different models of emotional intelligence capture different concepts. That is why, according to Locke (2005), emotional intelligence as a concept should be discarded. Cherniss (2010) disagreed with the statement that emotional intelligence as a concept should be "abolished." Based on his opinion, only one definition, the most accurate one, should be selected, while the others discarded. According to him, some criteria should be defined, based on which a definition can be linked to emotional intelligence or not. Other researchers, such as Bar-On (1997) and Petrides et al. (2007), believe that this problem is only temporary, so it is worth waiting to see whether other researchers find a solution to the problem in future (Kun, 2011; Cizrelioğulları & Babayiğit, 2022).

Methodology

Our research aimed to assess the impact of emotional intelligence on knowledge transfer among employees working in SMEs operating in the tourism sector and services. We carried out a questionnaire survey in April and May 2022. During our research, a database of thousands of companies was set up with the help of Finstat.SK website. The website allowed us to filter out companies operating in the tourism sector and services based on company size. We also got access to the email addresses of companies. The online questionnaire was created in Google Forms.

In some cases, the emails were incorrect; the recipient's mailbox could no longer receive emails; some company addresses were wrong, since they had changed names or owners, and some companies ceased by the time the emails were delivered. Since we were interested in the opinions of employees, we had to exclude responses from business owners in this research. Those companies, which do not represent the SME sector, were also excluded from this research. We managed to contact 500 companies, 315 employee responses were received, and 309 responses were processed.

Among the scales used to measure the level of emotional intelligence, one of the most well-known can be linked to Schutte et al. (1998). Assessing Emotions Scale (AES), as the authors refer to it (Schutte et al., 2009), was applied. The names SS-REI, SSEIT, EIS or SSRIE can also be used, but this has no particular significance in the operation of the method. The AES is based on a questionnaire consisting of 33 statements. The respondent completing the questionnaire has to respond to each statement on a 5-point Likert scale. The method is based on a three-pronged model developed by Salovey and Mayer (1990).

In contrast to the model, Schutte et al. (1998) classified 33 statements under a single factor when measuring emotional intelligence. This is called the global emotional intelligence (Kun et al., 2011; Gong & Paulson, 2016)

The following hypotheses were formulated in the research:

Hypothesis 1: Gender affects the level of EQ.

H0: There is no relationship between gender and EQ.
H1: There is a relationship between gender and EQ.

Hypothesis 2: The willingness of individuals to share knowledge depends on the level of emotional intelligence.

H0: The level of emotional intelligence does not influence the willingness to share knowledge.

H1: The level of emotional intelligence affects the willingness to share knowledge.

Pearson's chi-square test was used to examine the hypotheses:

$$\chi^2 = \sum_{total\ cells} \frac{(f_0 - f_e)^2}{f_e}$$

The relationship is primarily based on the comparison of observed (f0) and expected (fe) values.

In addition, we performed a Fisher's exact test in one case:

$$p = \frac{(a+b)!(c+d)!(a+c)!(b+d)!}{a!b!c!d!n!}$$

"a," "b," "c," "d" = individual frequencies
"N" = total frequency.

Cramer V indicator was used in one case to prove the strength of the relationship. N refers to the size of the sample, and k expresses the number of categories:

$$V = \sqrt{\frac{\chi^2}{N(k-1)}}$$

When testing the hypotheses, we used the generally accepted 5% margin of error ($\alpha = 0.05$). We used SPSS software to perform the tests.

We received a total of 309 responses to our questionnaire. The age groups were created based on Berkup's (2014) generational grouping. We made a small modification, as the youngest members of Generation Z were born in 2009. Individuals born after 2010 are already the members of Alpha generation (McCrindle et al., 2021). It is also important to highlight that we did not receive responses from individuals completing primary education.

Results

Determination of emotional intelligence

In order to obtain the level of emotional intelligence of the respondents, we used the Emotion Measurement Scale – AES (known as SSREI).

TABLE 2.1 Demographic characteristics of the respondents

Gender	Female	56.3%
	Male	43.7%
Age group	1965–1979	20.7%
	1980–1994	50.2%
	1995–2003	29.1%
Qualification	Secondary	51.1%
	Higher	48.9%
Monthly net income	500€ and below	13.5%
	501–1000€	46.1%
	1001–1500€	24.1%
	1501–2000€	10.5%
	2000€ and above	5.8%
Sector	Tourism	58.1%
	Services	41.9%

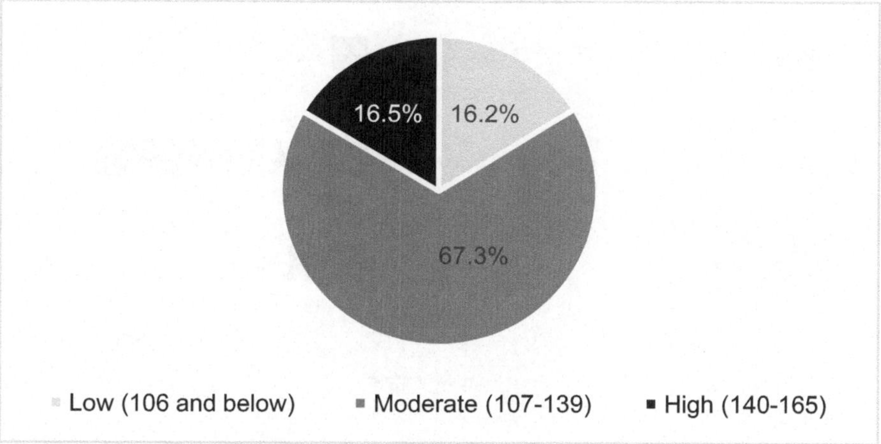

FIGURE 2.1 Applicant categories based on EQ

Since we did not want to draw a more profound psychological conclusion, we only focused on the total value of the scale. Interpreting some of the scales measuring EQ is a complicated task and requires professional expertise. However, the AES scale allows us to interpret it according to one factor, and its final result is a specific value. We can already use this value for comparison.

Schutte et al. (2009) also examined several research outputs before determining the applicant categories (low – medium – high), taking into account the standard deviation and the mean. We acted accordingly – the mean value was 123, while the standard deviation was 16. According to this, a value less than 107 is considered low, while a value above 139 is considered high.

Testing hypotheses

Hypothesis 1: Gender affects the level of EQ.

H0: There is no relationship between gender and EQ.
H1: There is a relationship between gender and EQ.

When testing Hypothesis 1, we wanted to know whether the level of emotional intelligence (EQ) in our sample shows similarities to Saklofske et al. (2007), which stated that women achieved higher EQ on average. We measured EQ by using the AES scale, which was an integral part of our questionnaire. However, this was divided into three categories based on the average and standard deviation as described earlier (Figure 2.1). Thus, our dependent variable is not metric – it is an ordinal type of variable. Based on this, we conducted a cross-table analysis and performed a chi-square test.

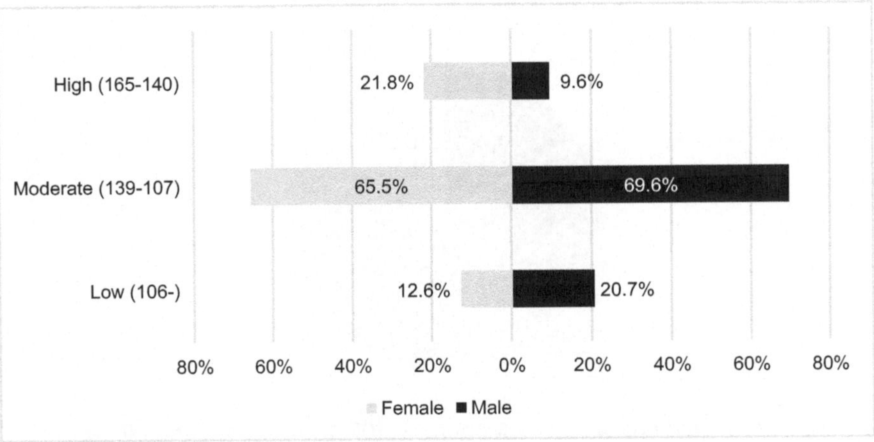

FIGURE 2.2 Relationship between gender and EQ of the respondents

TABLE 2.2 Distribution of EQ by gender

EQ (dependent)	*Gender (independent)*		
	Female	*Male*	*Total (pers.)*
High (165–140)	21.8%	9.6%	**51**
Moderate (139–107)	65.5%	69.6%	**208**
Low (106–)	12.6%	20.7%	**50**
Total (pers.)	**174**	**135**	**309**
Mean	125	121	123
St. dev.	16	16	16
Median	127	121	124
Mode	130	119	130

Table 2.2 facilitates a detailed understanding of the values. It shows the mean, standard deviation, median and mode values in addition to proportion values. Although the AES scale is based on a Likert scale, the authors of the scale (Schutte et al., 2009) worked also with the values as metric ones.

It can be seen that the most significant proportion of the respondents (both male and female respondents) have medium (average) EQ. In the case of women, there are a higher proportion of above-average performers, while in the case of men there are below-average performers. In the evaluation of the mean, median and mode values, a difference can be seen between the representatives of male and female respondents. Both in the case of male and female respondents, the standard deviation is the same as the standard deviation of the entire sample (16).

We wanted to perform a chi-square test for crosstab analysis. In order to achieve this, one more condition has to be met. In the case of the expected values, the value can be less than 5 in a maximum of 20 per cent in each field. To check this, we created the cross-table (Table 2.3), which shows both the observed ("Count" based on SPSS) and the expected values ("Expected count").

Based on Table 2.3, the condition to conduct the test is not violated, and the value is not less than 5. This is how the values in Table 2.4 can be interpreted:

The value of Chi is 10.137, which is higher than 5.991 corresponding to the given significance ($\alpha = 0.05$) and 2 degrees of freedom. So, the obtained value of the trial function is higher than the critical value of the chi distribution ($\chi2 > \chi2crit.$). Furthermore, the empirical significance level ($p = 0.006$) is also lower than the significance determined during our research ($\alpha = 0.05$). Based on the results, we can reject the null hypothesis. Since there is a significant relationship between our two variables, we can analyze the strength of this relationship.

In our case, the value of Cramer's V is 0.181, which is considered a moderate value. The value of indicator refers to the strength of the correlation between the variables, and it does not refer to their direction. Based on the research results, it can be assumed that women have higher EQ than men.

TABLE 2.3 Crosstab with observed and expected values – gender and EQ

			Gender (independent)		*Total*
			Female	*Male*	
EQ level (dependent)	Low	Count	22	28	50
		Expected Count	28.2	21.8	
	Moderate	Count	114	94	208
		Expected Count	117.1	90.9	
	High	Count	38	13	51
		Expected Count	28.7	22.3	
Total			174	135	309.0

TABLE 2.4 Chi-square test result – gender and EQ

	Value	*df*	*Asymp. Sig. (2-sided)*
Pearson Chi-Square	10.137	2	0.006
Likelihood Ratio	10.512	2	0.005
Linear-by-Linear Association	9.560	1	0.002
N of Valid Cases	309		

TABLE 2.5 Strength of the relationship – gender and EQ

	Value	*Asymp. Std. Error*	*Approx. T*	*Approx. Sig.*
Phi	0.181			0.006
Cramer's V	**0.181**			0.006
Contingency Coefficient	0.178			0.006
Gamma	0.342	0.102	3.208	0.001
N of Valid Cases	309			

Hypothesis 2: The willingness of individuals to share knowledge depends on the level of emotional intelligence.

H0: The level of emotional intelligence does not influence the willingness to share knowledge.

H1: The level of emotional intelligence affects the willingness to share knowledge.

We also examined whether individuals with higher emotional intelligence prefer to share their knowledge or not. Our dependent variable (willingness to share knowledge) was of binary and nominal types. The level of emotional intelligence was our independent variable (ordinal). We conducted a cross-table analysis and performed a chi-square test.

In the case of all defined categories of EQ, 100 per cent was still represented by the sum of the responses from the given group (Low n = 50, Moderate n = 208, High n = 51). In this case, we did not calculate the mean, standard deviation, mode and median values since the EQ categories (Low, Moderate, High) had been created based on the EQ value anyway. Thus, for example, the average of elements in "Low" category would naturally be lower than those in "High" category. Moreover, our dependent category is nominal and binary, which cannot be quantified, so we could not perform calculations (mean, standard deviation, mode, median).

We wanted to perform a chi-square test for the crosstab analysis. In order to perform analysis, one more condition had to be met, in the case of the expected values. The value can be less than 5 in a maximum of 20 per cent in each field. To check this, we created the cross-table (Table 2.6) presented here, which shows both the observed ("Count" based on SPSS) and the expected ("Expected count") values.

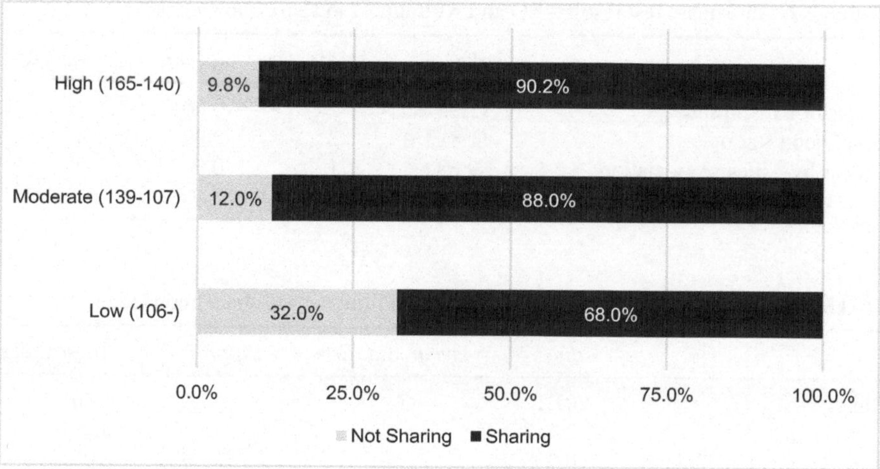

FIGURE 2.3 The relationship between EQ and willingness to share knowledge

TABLE 2.6 Crosstab with observed and expected values – EQ and willingness to share knowledge

			EQ level (independent)			*Total*
			Low	*Moderate*	*High*	
Knowledge sharing (dependent)	Not sharing	Count	16	25	5	46
		Expected Count	7.4	31.0	7.6	
	Sharing	Count	34	183	46	263
		Expected Count	42.6	177.0	43.4	
Total			50	208	51	309.0

Based on Table 2.6, the condition to conduct the test is fulfilled, and the value is not lower than 5. Therefore, the values in Table 2.7 can be evaluated:

The chi value is 13.947, which is higher than 5.991, corresponding to the given significance ($\alpha = 0.05$) and 2 degrees of freedom. So, the obtained value of the trial function is greater than the critical value of chi distribution ($\chi 2 > \chi 2crit.$). Furthermore, the empirical significance level ($p = 0.001$) is also lower than the significance determined during our research ($\alpha = 0.05$). Based on the results, we can reject the null hypothesis. Since there is a significant relationship between our two variables, the strength of the relationship can be tested. We used Cramer's V again.

In our case, the value of Cramer's V is 0.212, which is a moderate value. However, the value only indicates the strength of the correlation between the variables and not their direction. The obtained results show that people with higher emotional intelligence are more open to knowledge sharing within the companies.

TABLE 2.7 Chi-square test result – EQ and willingness to share knowledge

	Value	*df*	*Asymp. Sig. (2-sided)*
Pearson Chi-Square	13.947	2	0.001
Likelihood Ratio	11.813	2	0.003
Linear-by-Linear Association	9.682	1	0.002
N of Valid Cases	309		

TABLE 2.8 Strength of the relationship – EQ and willingness to share knowledge

	Value	*Asymp. Std. Error*	*Approx. T*	*Approx. Sig.*
Phi	0.212			0.001
Cramer's V	**0.212**			0.001
Contingency Coefficient	0.208			0.001
Gamma	0.427	0.137	2.744	0.006
N of Valid Cases	309			

Conclusion

If there is direct interaction with customers (e.g. any job related to tourism, customer service, Bartók et al., 2021), other services), or when the duties to be performed require teamwork and cooperation within the company (e.g. between employees, or in the manager–subordinate relationship) or outside (e.g. with corporate partners), it is a benefit to employ people with high EQ. It is important to mention that not only subordinates but also middle managers and those in cooperation with subordinates have an adequate level of emotional intelligence.

Gender should be considered, when selecting employees for the appropriate position, but the principle of non-discrimination based on gender has to be respected. If the company is looking for an employee to fulfill a position, where knowledge sharing plays a significant role, it is vital to hire a candidate, who also has the appropriate skills and knowledge to enter the position.

The AES scale may be appropriate for measuring emotional intelligence since its evaluation does not require extensive knowledge in the field of psychology. The scale is relatively voluminous, but it is still one of the frequently used methods for measuring EQ. Its biggest advantage is that it quantifies the EQ of individuals. Based on this, the manager or the representatives of HR can compare the scores of the applicants, and give priority to the one(s) with the highest EQ score.

During our research, we had to face financial and time constraints. It is also worth highlighting the main criticism of the quantitative methods. The respondents completing the questionnaire interpreted the individual questions differently. Furthermore, these tests only show how individuals judge their own qualities or themselves; however, this may differ from the reality. It is also important to mention that our questionnaire turned out to be voluminous. It was mainly due to the AES scale. We did not obtain an adequate amount of answers from employees with basic school qualification. The number of employees with basic school qualification is relatively low, but they may not have received the questionnaire since their Internet access at the workplace is limited or do not have at all.

In future, we would like to reach more respondents and also use the opportunities provided by offline space. Furthermore, we would like to conduct interviews with the managers of SMEs operating in tourism sector and services to obtain an overall picture. We would also incorporate our experience gained during this research into the processes.

References

Ajide, F. M. (2022). Impact of tourism development on latent entrepreneurship in BRICS. *Entrepreneurial Business and Economics Review, 10*(4), 143–154.

Aydin, I. (2022). Effects of work-related stressors and work engagement on work stress: Healthcare managers' perspective. *Central European Business Review, 11*(4), 47–62.

Bar-On, R. (1997). *The Emotional Quotient inventory (EQ-i): A test of emotional intelligence*. Multi-Health Systems, Inc.

Bartók, O., Kozák, V., & Bauerová, R. (2021). Online grocery shopping: The customers′ perspective in the Czech Republic. *Equilibrium. Quarterly Journal of Economics and Economic Policy, 16*(3), 679–695. https://doi.org/10.24136/eq.2021.025.

Beldoch, M. (1964). Sensitivity to expression of emotional meaning in three modes of communication. In *The Communication of Emotional Meaning* (pp. 31–42). McGraw-Hill.

Berkup, S. (2014). Working with generations X and Y in generation Z period: Management of different generations in business life. *Mediterranean Journal of Social Sciences, 5*(19), 218–229.

Bozsik, N., Ngo, D. M., & Vasa, L. (2023). The effects of foreign direct investment on the performance of small-medium enterprises: The case of Vietnam. *Journal of International Studies, 16*(1), 57–70.

Brackett, M. A., & Mayer, J. D. (2003). Convergent, discriminant, and incremental validity of competing measures of emotional intelligence. *Personality and Social Psychology Bulletin, 29*(9), 1147–1158.

Cherniss, C. (2010). Emotional intelligence: Toward clarification of a concept. *Industrial and Organizational Psychology, 3*(2), 110–126.

Civelek, M., Ključnikov, A., Fialova, V., Folvarčná, A., & Stoch, M. (2021). How innovativeness of family-owned SMEs differ depending on their characteristics? *Equilibrium. Quarterly Journal of Economics and Economic Policy, 16*(2), 413–428. https://doi.org/10.24136/eq.2021.015.

Cizrelioğulları, M. N., & Babayiğit, M. V. (2022). Effects of high-performance work system on job satisfaction: The mediating role of psychological capital in the hotel employees of North Cyprus. *Journal of Tourism and Services, 24*(13), 43–70. doi: 10.29036/jots.v13i24.284

Davenport, T. H., & Prusak, L. (1998). *Working knowledge: How organizations manage what they know*. Harvard Business School Press.

Drucker, P. F. (2011). *Post-capitalist society*. Routledge.

Foss, N. J. (1997). Equilibrium vs evolution in the resource-based perspective: The conflict legacies of Demsetz and Penrose. *DRUID Working Paper*, *97*(10).

Garamvölgyi, J., & Rudnák, I. (2015). Kulturális intelligencia bemutatása, mérési lehetőségei. *Studia Mundi – Economica*, *2*(3), 113–126.

Gardner, H. (1983). *Frames of mind: The theory of multiple intelligences*. Basic Books.

Goleman, D. (1995). *Emotional intelligence*. Bantam Books.

Gong, X., & Paulson, S. E. (2016). Validation of the Schutte self-report emotional intelligence scale with American College Students. *Journal of Psychoeducational Assessment*, *36*(2), 175–181.

He, K., Oláh, J., & Hasan, M. (2022). The influence of psychological ownership and social support on organizational resilience: The mediating role of organizational identity. *Journal of Business Economics and Management*, *23*(3), 650–667.

Hodgson, G. M. (1998a). Evolutionary and competence-based theories of the firm. *Journal of Economic Studies*, *25*(1), 25–56.

Hodgson, G. M. (1998b). Competence and contract in the theory of the firm. *Journal of Economic Behavior and Organization*, *35*(2), 179–201.

Kapás, J. (1999). Egy új vállalatelmélet: erőforrásalapú megközelítés. Összefoglalás és diagnózis. *Vezetéstudomány*, *30*(2), 35–43.

Ključnikov, A., Civelek, M., Fialova, V., & Folvarčná, A. (2021). Organizational, local, and global innovativeness of family-owned SMEs depending on firm-individual level characteristics: Evidence from the Czech Republic. *Equilibrium. Quarterly Journal of Economics and Economic Policy*, *16*(1), 169–184. https://doi.org/10.24136/eq.2021.006.

Ključnikov, A., Civelek, M., Klimeš, C., & Farana, R. (2022). Export risk perceptions of SMEs in selected Visegrad countries. *Equilibrium. Quarterly Journal of Economics and Economic Policy*, *17*(1), 173–190. doi: 10.24136/eq.2022.007

Kumar, S., Valeri, M., & Shekhar (2021). Understanding the relationship among factors influencing rural tourism: A hierarchical approach. *Journal of Organizational Change Management*, *35*(2), 385–407. doi: 10.1108/JOCM-01-2021-0006

Kun, B. (2011). *Az érzelmi intelligencia és az emocionális és szociális kompetenciák szerepe a pszichoaktívszer-használatban*. ELTE.

Kun, B., Urbán, R., Balázs, H., Kapitány, M., Nagy, H., Oláh, A., & Demetrovics, Z. S. (2011). Az érzelmek mérése skála háromfaktoros modelljének adaptálása = Factor structure of the Hungarian adaptation of Assessing Emotions Scale. *Magyar Pszichológiai Szemle*, *66*(3), 449–466.

Lane, R. D., & Schwartz, E. M. (1987). Levels of emotional awareness: A cognitive-developmental theory and its application to psychopathology. *American Journal of Psychiatry*, *144*(4), 133–143.

Leuner, B. (1966). Emotional intelligence and emancipation. *Praxis der Kinderpsychologie und Kinderpsychiatrie*, *15*(1), 193–203.

Li, Y., Ren, N., & Chaudhry, S. S. (2008). An enterprise knowledge management system based on the use case model. In *II. International Conference on Research and Practical Issues of Enterprise Information Systems (CONFENIS 2007)* (pp. 1141–1146). Springer.

Locke, E. A. (2005). Why emotional intelligence is an invalid concept. *Journal of Organizational Behavior*, *26*(4), 425–431.

Mayer, J. D. (2015). *Personal Intelligence*. New York: Farrar, Straus and Giroux.

McCrindle, M., Fell, A., & Buckerfield, S. (2021). *Generation alpha*. Hachette Australia.

Michie, J. (2019). Edith Penrose's contribution to real world economics: Introduction. *International Review of Applied Economics*, *33*(5), 599–600.

Milczarek, A., Miłaszewicz, D., Nagaj, R., Szkudlarek, P., & Zakrzewska, M. (2015). Social networks as a determinant of the socialisation of human capital. *Human Resources Management & Ergonomics*, *9*(2), 89–103.
Neale, S., Spencer-Arnell, L., & Wilson, L. (2009). *Emotional intelligence coaching*. Kogan Page.
Payne, W. L. (1986). *A study of emotion: Developing emotional intelligence; self-integration; relating to fear, pain and desire*. University Microfilms.
Penrose, E. (1959). *The theory of the growth of the firm*. Blackwell.
Perovic, J., Raicevic, M., & Delibasic, M. (2021). Foreign direct investment, tourism, economic growth and trade: Panel Vector Auto Regression (Var) Model. *Transformations in Business & Economics*, *20*(1), 140–153.
Petrides, K. V., Pita, R., & Kokkinaki, F. (2007). The location of trait emotional intelligence in personality factor space. *British Journal of Psychology*, *98*(2), 273–289.
Rumelt, R. (1984). Towards a strategic theory of the firm. *Competitive Strategic Management*, *26*, 556–570.
Saarni, C. (1990). Emotional competence: How emotions and relationships become integrated. In *Socioemotional development: Nebraska symposium on motivation* (pp. 115–182). University of Nebraska Press.
Saklofske, D. H., Austin, E. J., Galloway, J., & Davidson, K. (2007). Individual difference correlates of health-related behaviours: Preliminary evidence for links between emotional intelligence and coping. *Personality and Individual Differences*, *42*(2), 491–502.
Salovey, P., & Mayer, J. D. (1990). Emotional intelligence. *Imagination, Cognition and Personality*, *9*(3), 185–211.
Santos, V., Sousa, B., Ramos, P., & Valeri, M. (2022). Emotions and involvement in tourism settings. *Current Issues in Tourism*, *25*(10), 1526–1531. doi: 10.1080/13683500.2021.1932769
Sarihasan, I., Dajnoki, K., Oláh, J., & Al-Dalahmeh, M. (2022). The importance of the leadership functions of a high-reliability health care organization in managing the Covid-19 pandemic in Turkey. *Economics and Sociology*, *15*(1), 78–93.
Schutte, N. S., Malouff, J. M., & Bhullar, N. (2009). The assessing emotions scale. In *Assessing emotional intelligence: Theory, research, and applications* (pp. 119–134). Springer International Publishing.
Schutte, N. S., Malouff, J. M., Hall, L. E., Haggerty, D. J., Cooper, J. T., Golden, C. J., & Dornheim, L. (1998). Development and validation of a measure of emotional intelligence. *Personality and Individual Differences*, *25*(2), 167–177.
Štefko, R., Jenčová, S., & Vašaničová, P. (2020). The Slovak spa industry and spa companies: Financial and economic situation. *Journal of Tourism and Services*, *20*(11), 28–43. doi:10.29036/jots.v11i20.137
Sternberg, R. J. (2000). *Handbook of intelligence*. Cambridge University Press.
Strugar Jelača, M., Bjekić, R., Berber, N., Aleksić, M., Slavić, A., & Marić, S. (2022). Impact of managers' emotional competencies on organizational performance. *Sustainability*, *14*(14), 8800. doi: 10.3390/su14148800
Szabó-Szentgróti, G., Gelencsér, M., Szabó-Szentgróti, E., & Berke, S. (2019). Generációs hatás a munkahelyi konfliktusokban. *Vezetéstudomány – Budapest Management Review*, *50*(4), 77–88. doi: 10.14267/VEZTUD.2019.04.08
Thorndike, E. L. (1920). A constant error in psychological ratings. *Journal of Applied Psychology*, *4*(1), 25–29.
Tovmasyan, G. (2023). Factors that influence domestic tourism demand: Evidence from Armenia. *Economics and Sociology*, *16*(2), 75–88.

Vágnerová, M. (2005). *Základy psychologie*. Karolinum.

Valeri, M., & Baggio, R. (2021). Increasing the efficiency of knowledge transfer in an Italian tourism system: A network approach. *Current Issues in Tourism*, *25*(13), 2127–2142. doi: 10.1080/13683500.2021.1937960

Wach, K., Maciejewski, M., & Głodowska, A. (2023). Inside entrepreneurial orientation: Do risk-taking and innovativeness influence proactiveness? *Economics and Sociology*, *16*(1), 159–175.

Wechsler, D. (1955). *Manual for the Wechsler adult intelligence scale*. Psychological Corporation.

Wernerfelt, B. A. (1984). Resource-based view of the firm. *Strategic Management Journal*, *5*(2), 171–180. doi: 10.1002/smj.4250050207

Wierzbiński, B., Zaborek, P., Wosiek, M., & Surmacz, T. (2023). Knowledge management orientation as a driver of competitive performance: Evidence from Polish SMEs in the aviation industry. *Entrepreneurial Business and Economics Review*, *11*(1), 93–112.

Yin, J., Cheng, Y., & Wang, K. (2020). Deconstructing the tourism cooperation network in the Northeast Asia Region: Characteristics and attributes. *Transformations in Business & Economics*, *19*(2B) (50B), 783–807.

Zakrzewska, M., Jarosz, S., Piwowar-Sulej, K., & Sołtysik, M. (2022). Enterprise agility – its meaning, managerial expectations and barriers to implementation – a survey of three countries. *Journal of Organizational Change Management*, *35*(3), 488–510. doi: 10.1108/JOCM-02-2021-0061

Zamir, Z. (2019). The impact of knowledge capture and knowledge sharing on learning, adaptability, job satisfaction and staying intention. *International Journal of Entrepreneurial Knowledge*, *7*(1), 46–64. doi: 10.2478/ijek-2019-0004

3

MANAGING THE USE OF VERBAL AND NON-VERBAL COMMUNICATION IN TOURISM

Stylianos Bouzis, Panoraia Poulaki and Marco Valeri

Introduction

The tourism environment, in which tourism communication takes place, consists of words, images, and sounds. In addition, specific channels of communication may appear, which are logically subject to a specific syntactic morphology. Ethnographic communication is therefore underlying tourism discourses.

The rapid growth of tourism has resulted in its introduction into the scientific literature. The fact that many sciences are related and applied to tourism leads to the introduction and use of specific terms or concepts. This terminology is not fully embedded in the minds of the common tourist, scholar, reader or interested person. Therefore, in the first phase, following this chapter, the delineation of the content of these terms, which will be encountered several times in the rest of the research, is carried out.

In communication between tourists and between tourists and locals, various channels of communication are often used. The organization of discourse is often characterized by certain defined rituals and consequently by functional varieties. These functional varieties present utilitarian differences which derive from the realization of discourse in different communicative situations.

The communicative process is divided into verbal and non-verbal parts. This separation is carried out according to the communication code that the actors involved have chosen to use. Usually, the type of message and therefore the communication method are chosen by the sender as he or she defines the initiation and the "rules" of communication. Etymologically, the term "communication" comes from the Latin language, in particular the word "communis" which translates as "common."

DOI: 10.4324/9781032633374-4

The non-verbal elements that accompany communication are key features of non-verbal communication. Enriching the communication process with such elements makes a decisive contribution and is effective. So how is non-verbal communication defined and what does it involve?

As for the importance of this research in the tourism sector, it contributes to more effective communication between tourists and locals, through the need for the coexistence of verbal and non-verbal communication (as demonstrated by the results of the primary research). In addition, the research lies in highlighting the frequent use of non-verbal communication practices without them overshadowing the verbal behavior between tourists and locals. Finally, the achievements presented by this research effort are particularly noticeable in the disciplines of language, communication, and tourism, as their effects are felt in all relevant fields of reference.

Tourism and its relationship with communication

First, it is important to clarify the concept of "tourism," a seemingly easy but extremely multi-dimensional term. Tourism is the activity that develops when people travel and includes both the preparation and the realization of the trip with the stay, the return, and the memories made. In addition, it includes the activities during the trip, shopping, and the interactions of tourists with local people. It is mentioned by Lickorish and Jenkins (2004) that tourism has been a mass phenomenon since 1950. However, according to UNWTO surveys, in recent decades tourism has experienced continuous growth and is one of the fastest-growing economic sectors.

On the other hand, a tourist is a social category of person in which the socio-economic characteristics of those who travel within or outside their country and are motivated by any reason other than paid work are observed (Tsartas, 1996). At this point, the definition of "tourist" is clarified, as in the empirical context of the research it will be commented on in contrast to residents and the communicative linguistic and non-linguistic relationships they establish during the trip.

Sociolinguistics

This is the newly emerging discipline that emerged around the mid-1960s and is concerned with human language and communication and the relationship it enters with social norms. Speaking specifically of linguistics, it studies the structure of natural languages and modern linguistic facts (Heller, Pujolar & Duchêne, 2014). The description of linguistics in this research is at the same time related to sociology since verbal communication is a social and not a purely linguistic phenomenon. The main research areas of study in sociolinguistics are as follows:

- Social identity of the speaker
- Social identity of the listener

- Social environment in which verbal communication takes place
- Different social and linguistic behaviors of speakers

Communication

Communication per se deals with the transmission and exchange of information, positions, opinions, and ideas (Lytras, 2008) and is a fundamental element of the human cognitive framework. In other words, speaking of communication means the process of transmitting a message with the aim of mutual understanding between the transmitter and the receiver. Individuals who communicate often enter dependent relationships. The following are presented as tools for carrying out communication:

- Oral speech
- Written language
- Image
- Expressions and gestures
- Sound
- Combinations of the above

Communication as a phenomenon applies to all sciences and is the subject of interdisciplinary approaches as it takes place wherever the human factor is encountered. The term communication first appears in the works of Plato and Aristotle, where it is shown that all social processes presuppose the existence of some form of communication.

The relationship between the key concepts

At the beginning of the first decade of the twenty-first century, the language of tourism began to flourish as a field of study and has been the subject of research in many areas related to tourism (e.g. marketing, branding, advertising, sociology of tourism). Regarding the latter area, the attention of researchers is focused on the relationship between language and society, which as a subject of study is characterized as quite multifaceted, as it is analyzed through various perspectives. The specific research at stake is the following: "*How do social structures influence the linguistic behavior of tourists?*" It is worth mentioning that in a social situation such as tourism, specific vocabulary is used in all possible tourist contexts. When using language, it is possible to see how tourists and locals use the variability of language in pursuit of personal and professional goals. Tourist discourse derives from foreign discourse and is usually in favor of the tourist, as it demands that local inhabitants of a lower social level know the language spoken by the tourist. Moreover, the host language has been shown to be an apathetic element for both mass tourists and individual tourists since it seems to be used only in specific cases of travelers – mainly by people traveling with the goal of exploring landscapes, customs, cultures, and tourist destinations in general.

Communication in the tourism industry is presented in the form of intercultural communication, in which various social and linguistic features are underlying. A local resident, for example, is required to communicate with tourists through specific linguistic features, since they come from different social and cultural backgrounds. Usually, this communication takes place in English. However, the cultural background and values of the tourists who encounter the local resident must be understood before the local resident can determine the linguistic features that will accompany their communication. In conclusion, the linguistic variations that take place in tourist communication are primarily influenced by the social characteristics and culture of the interlocutors.

Verbal communication

Verbal communication is the mainstay of oral communication, in which the message is spread through the spoken words and between speakers. Verbal communication expresses in words the emotions and externalizes the thoughts and ideas expressed by the sender while transmitting the information to the receiver. In addition to the above, it includes sounds, words, and speech, the latter being the most effective way of communication. C. Babiniotis (2022) claims that "The linguistic code consists of language signs and words that aid verbal communication." As a process, verbal communication involves seven (7) key points, as shown in Figure 3.1. In this process, the sender initiates the conversation by encoding his or her thoughts with specific words to convey a message (written or spoken). For the message to be effective from the receiver's point of view, the communication channel must be carefully chosen so that the receiver can fully understand it and then decode it in the best possible way. Feedback is the final stage in which the effectiveness of the message is communicated to the sender. At this point, it is worth noting that the recipient's response may be verbal or non-verbal. Finally, noise is defined as all obstacles to communication caused by misinterpretation.

Types of verbal communication in tourism

Verbal communication, as it develops in the tourism sector, branches into four types (according to Littlemore, 2003). Firstly, interpersonal communication occurs most frequently in the tourism process. It is conducted between two persons and involves the dialogue formed by the transmitter and receiver. For the communication to evolve and become clearer, the two speakers exchange the status of the respondent and interviewee during the communication. A typical example is the dialogue at the reception of a hotel or in a tourist shop between a tourist and an employee. A second form is intrapersonal communication, and this is a private communication, which is limited to the tourist's inner life. This includes silent conversations, which

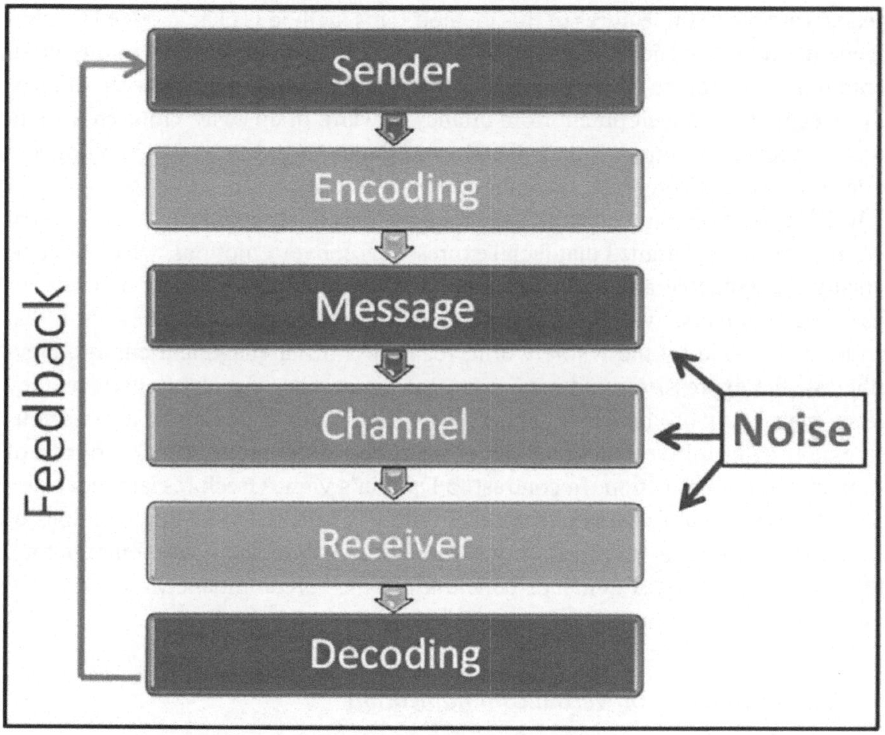

FIGURE 3.1 Seven major elements of communication process

usually elaborate on thoughts, actions, and choices. In tourism, this type of communication occurs during the preparation and planning of the trip, where the person concerned listens internally to the information provided about the destination.

When there are more than two people involved in the conversation, the small communication group type is identified. The number of people involved may be small, as there is interaction and conversation during the process. An example of group communication in tourism is a tour of an archaeological site, during which the visitors could talk to each other or to the guide if they are accompanied by the guide. At the same time, there should always be a specific purpose of setting up a small communication group, so that no chaos is caused by the entrance of more involved speakers. Finally, public communication is rarely found in tourism, as it studies the communication process between an individual and a large gathering of people. In this case, the transmitter is the individual who speaks alone, and the public has the role of the receiver.

Non-verbal communication

Non-verbal communication externalizes the feelings and thoughts of the speakers during communication. The posture, facial expressions, and gestures of the speakers are the main features of this method. This method is characterized by phylogenetic and ontogenetic priority, which means the pre-existence of non-verbal communication before verbal communication. More specifically, non-verbal techniques begin their development from infancy onward. In this way, children learn to greet someone, and through their facial expressions or gestures, their willingness to do something is perceived since they cannot yet use words.

In 1872, Darwin, an American psychologist, was the first to refer to non-verbal communication and argued that facial expressions convey emotions, which are universally recognizable and serve an important communicative function. They appear to be of universal value, since smiling, crying, and caressing are behaviors that can be understood by the whole world, regardless of language and culture. Nevertheless, the expressions and movements of the speaker are involuntary, as they are spontaneously produced by the nervous system. This is due to the unconscious regression of primitive reactions and is justified by the phylogenetic priority of non-verbal communication. In contrast to Darwin's view, Greek researchers claim that with non-verbal features one speaker influences the behavior and emotions of the other. In any case, opinions on the weight of each of them are contradictory, although the majority of evidence concludes in the "predominance" of nonverbal over verbal communication.

Its relationship with non-verbal communication

Although these two methods of communication are important and equally effective, the question arises as to the actual relationship that develops between them. Verbal communication, with the phrases and expressions it uses, cannot de facto be fully effective without the help of "psychic filters." In other words, although one complements the other, non-verbal communication seems to indicate a possible predominance as it accounts for more than 65% of all messages exchanged. This fact is justified by the emission of messages and expressions not only during the conversation but even when the speakers are not speaking. Consequently, non-verbal techniques occur both during communication and when there is a pause. It is also stated that meaning in linguistic communication is not given by the grammar and syntax of the spoken message but by the channels of non-verbal communication. Worthy of mention is Freud's assertion that it is impossible for speakers to keep a feeling or a secret hidden since it is easily and unconsciously betrayed many times by non-verbal behavior. Finally, it is concluded that there is a healthy competition between them. When non-verbal methods of communication conflict with what the speakers say, they are contradicted by what is perceived by attitude and expression.

Features of non-verbal communication in tourism

The language system includes the motor and expressive activities that flow from the speaker's body. As a method, it incorporates many different channels of communication, which are independent of the verbal or any other linguistic element. The features that externalize elements of speakers and are embodied in body language are distinguished by certain sub-elements (elements) depending on the sensory organ through which they are manifested. These characteristics are presented as follows:

General appearance: 3 (of the local inhabitants and especially of the workers) in businesses surrounded by tourists. It determines the tourist's first impression of the destination and its people. Clothing in combination with the personality, behavior, age, and gender of the locals acts as information for the tourist. Also, the general appearance is the beginning of the relationship between tourists and locals, as it is the "self-disclosure" of the local speaker.

Eye contact: The speaker's gaze is considered the most important non-verbal element of communication, as it is the first step in the beginning of an interpersonal conversation and promises to express emotions. In the tourism sector, e.g., in a hotel, when the receptionist manages to make eye contact with tourists, the achievement of his/her objectives (in this case, to become friendly and cheerful) begins. It is often observed that the employee changes his or her gaze and does not look at the tourist but at some point, in the room. This is usually due to a lack of confidence, stress, and an inability to meet the demands of tourists. Finally, "snitching with the eyes" is mainly done by locals who want to impose discipline and is accompanied by similar expressions.

Facial expressions: It has been proven that this body language technique makes it extremely difficult for researchers to decode and therefore interpret. This is primarily because the expression of emotions through the face lasts only a few seconds and, secondarily, because of the complexity and variety of these facial expressions. Nevertheless, the most spontaneous and frequent expression is the smile, internationally recognized and associated with intimacy and sympathy.

Hand movements and postures (gestures): These are used by the local resident to be better understood by the tourist. This gives emphasis and vividness to what is said. A typical example of such an interpretation is the crossing of hands or hands in pockets. When a tourist asks a local passerby for information about going to a monument and the local has his hands in his pockets, this shows carelessness and lack of seriousness. Crossed hands, on the other hand, indicate a dividing line between tourists and locals. Regarding the overall posture and taking again as an example the communication between a hotel employee and a tourist, in the case of a mistake in the latter's posture, it is very common to lose eye contact between the two speakers. This can happen when the receptionist bows his head and concentrates on the documents in front of him or on the computer. This can cause problems. How can he avoid it? By not turning his whole body and gaze away from his interlocutor, but by keeping an intermediate position (of the ¾ type) in which he can make eye contact with both the tourist and his occupation.

Effectiveness of non-verbal communication

Having reached the end of the chapter and having covered all the theoretical aspects regarding the basic concepts of the research, the effectiveness rate of non-verbal communication starts to be discussed and reinforced. In earlier and highly representative research carried out in the late 1960s, Albert Mehrabian demonstrates in a practical way that non-linguistic elements are most often found in a communication process. Mehrabian creates a computational formula that studies how verbal elements match non-verbal elements. He applies it specifically to identify communicative techniques and channels in a message coming from a particular speaker because of his body language. The following graph depicts the elements of personal communication, according to the findings of the researcher. It is immediately apparent that there is an overwhelming predominance of non-verbal characteristics, of which 55% concern elements of body language (e.g. eye contact, expressions, gestures) and 38% phonological elements that study the tone and style of voice (Figure 3.2). However, spoken words account for only 7% of communication. It is ultimately concluded that non-verbal communication is considered more reliable than verbal communication. This is even reinforced when verbal and non-verbal messages are confronted.

Methodology

The primary data in this case were collected by sharing questionnaires. Therefore, to achieve the objective 202 questionnaires were distributed to draw conclusions and formulate recommendations and solutions. The ultimate purpose of this quantitative research is to test the following research hypotheses:

1 Which method of communication is superior during travel preparation?
2 How is the communication practice of locals towards tourists outlined?
3 How do tourists manage verbal communication?
4 Which method is most used when tourists and locals interact?

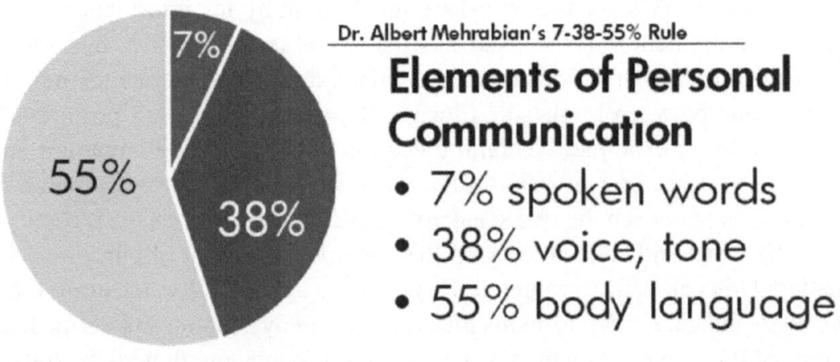

FIGURE 3.2 Albert Mehrabian's 7–38–55 Rule of Personal Communication

The individual grouped questions stem from these questions.

In terms of the type of research method, the following research is characterized as primary, quantitative, and interdisciplinary. Its interdisciplinary nature is evidenced by the fact that the involved disciplines encountered in the research are three (3): language, communication, and tourism. In addition, it is partly descriptive in terms of its intended purpose. The research attempts to explain a situation; therefore, it is rightly considered ethnographic with interpretative/qualitative approaches, especially in the theoretical part it examines.

Coming to the point of analyzing the stages of the methodology followed by the research, the researcher aims to test the outcome of his/her hypotheses with the help of the collected data. A questionnaire is the main instrument for collecting them.

Survey technique and sampling

The sampling method chosen to collect the sample necessary for the research is accessibility or convenience sampling. It belongs to the group of non-probability sampling methods, in which the probability of selection of everyone in the sample in the study population is unknown (Galanis, 2017). It is perhaps one of the most widespread and used forms of sampling. Samples of this method are selected from the population based on convenience, rather than randomness. This gives the opportunity to anyone who wishes to do so to take part in the survey and answer the research questions. Another method of collecting responses is presented by the distribution of the questions in a central and fixed place where passers-by could participate in the survey (Floyd & Fowler, 2001).

Sampling empathizes with the issue of representativeness and takes care to conduct inferences that can be generalized to the whole population because of the large sample size. Useful material for conducting the theoretical part of the research are charts and studies of already published sources of information (secondary sources).

The population targeted by the survey comes from a specific geographical area and specifically from Greece. In general, however, there is no need to collect specific characteristics, due to the simplified and general nature of the survey. Therefore, there are no restrictions on the population of respondents, while the only common characteristic that should have been found among them is the fact that they all had to have traveled at least once; that is, they all had to be potential tourists, without any additional restriction other than their Greek identity.

Results

In this part, a check is carried out to answer the research questions, as they were posed from the beginning. Putting the demographic characteristics of the respondents on the table, it is recorded that there are 202 respondents, of which 138 are women and 64 are men. Their ages range from 15 to 71 years old with an average

age of 29.22 years. Most of them are graduates of higher education institutions (105 persons), while several hold a postgraduate degree (47), 22 are high school graduates, 25 are college graduates and three hold a doctoral degree; 74.8% travel rarely to not at all frequently, and the remaining 25.2% travel frequently to quite frequently. The majority prefer to stay in a hotel (64.9% of the sample), followed by Airbnb accommodation (14.9%), rooms to let (9.4%), a friend or relative's house (7.9%) and a small percentage in camping (1%).

The research question is "Which method of communication is preferred when preparing the trip?" According to respondents' answers, non-verbal communication is far superior to verbal communication. This conclusion is initially justified by the overwhelming difference recorded in the chosen means of destination information. Only 21 people, or 10.4%, said that they were informed about the destination they were going to visit by face-to-face communication with friends, relatives and/or tourist agencies. However, the majority (76.7%) say that they search on internet engines, while the rest mention social media, printed tourist guides and information from other types of websites (e.g. vlogs, blogs). In addition, most respondents (66%) believe that an admirable photo of the destination with only the basic information contributes more to the final choice, while the remaining 34% believe that a complete, detailed, and highly verbose description in words can make it more effective in attracting tourists (Figure 3.3).

It therefore becomes clear here too that non-verbal communication is superior since a photograph is a non-verbal practice and a written text (speech) is a verbal

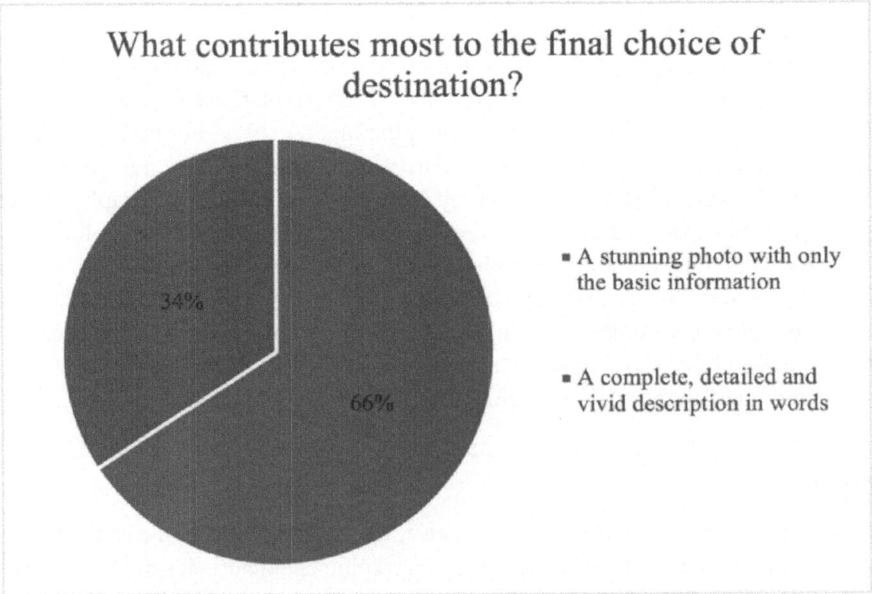

FIGURE 3.3 Pie chart of frequency variable "Contribution of choice"

feature of the language of advertising. In this respect, the research confirms the theoretical study of Reilly's (1988, p. 109) book, which claims that for the message to be as effective as possible, it must be short and clear. Similarly, regarding the image, most researchers argue that a photograph should be accompanied by accompanying text. This does not, however, negate the importance of visual promotional elements. It has been shown that 75% of tourist brochures consist of pictures (Dilley, 1986, p. 60) and this is a key indication that there is an emphasis on the importance of non-verbal and visual messages.

Regarding the question "How do you outline the communication practice of locals towards tourists?", 45.5% state that they rarely or not at all use gestures when talking to a local. Therefore, due to the minimal use of gestures, no elements of the personalities of the people involved are revealed. On the contrary, 64.9% emphasize the tone of voice when speaking. According to other research data, voice volume is innate, while low voice volume among other things is due to psychological reasons and conceals certain elements of character. When the listener tries to get the speech, louder speech volume is observed (Brettos, 2003, p. 109). Also, nonverbal communication including voice volume constitutes 93% of the spoken message, while only 7% is part of verbal communication.

The majority (87%) believe that the local resident utilizes nonverbal communication to develop a positive atmosphere between him/her and tourists (Figure 3.4).

According to Pease A. & B., looking, smiling, and shaking hands are very important in creating first impressions, "breaking the ice" and developing a positive atmosphere and familiarity. Nevertheless, 104 respondents to the survey believe

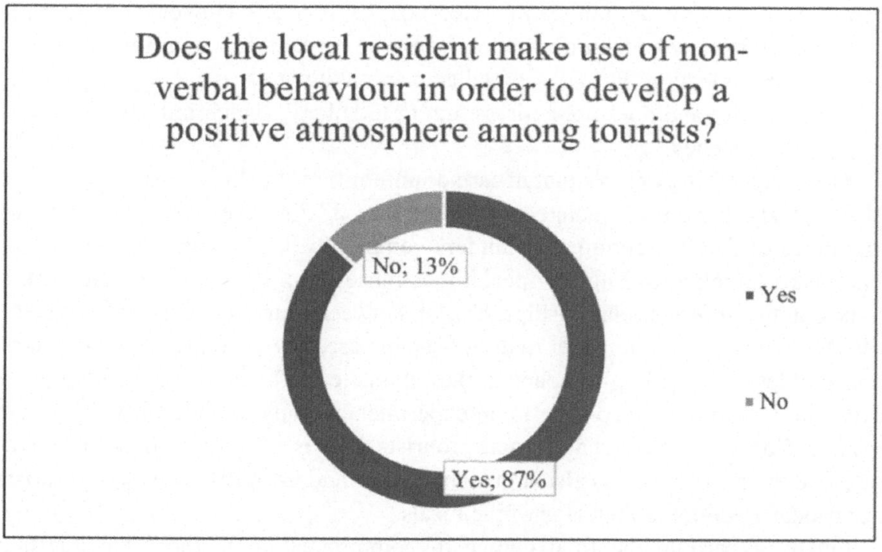

FIGURE 3.4 Pie chart of frequencies for utilization of non-verbal behavior by locals

that residents do not maintain contact with tourists after the holiday is over, while 98 believe they do. The difference is clearly not large, but it is worth noting that making friends and maintaining relationships depends on other situational factors.

Finally, 79.2% of the respondents claim that the native during their communication uses body movements, facial expressions, and dialogue at the same time. The enrichment of speech with various other non-verbal means, such as posture, can reveal information about the mood of both the speaker and the listener since it indicates the direction of interest and the need to abandon the conversation. In conclusion, it can be concluded that locals on their part use more often non-verbal communication practices, which in fact often frame and reinforce their words.

Regarding the question "How do tourists manage verbal communication?" 50% of the respondents state that they passively listen to their local interlocutor without interrupting him/her when he/she has the floor, 37% listen with interest, 8.4% remain apathetic, 3.5% enrich the conversation with comments and 1% listen with interest and address him/her in case a piece of information is not understood. The majority, therefore, showed abstention and/or apathy during the discussion (58.4%). This attitude is partly justified, since there is confusion of many stimuli/messages received by the tourist. The tourist is therefore in the process of processing so that only the useful information is listened to, and the meaningless data is automatically passed on to the subconscious. The 113 respondents believe that the dialogue between tourists and local people only occurs in certain circumstances (such as when buying tourist products) and not throughout the trip. This may of course be because tourists do not seek face-to-face communication and get information about whatever they want from the internet, as demonstrated in a previous finding of the survey. However, every social process requires communication since people cannot avoid communicating (Gotovos, 1990, p. 59). However, they may in certain circumstances seek information from sources that do not require the use of verbal communication. In total, 166 talks to an unknown native in the singular for more familiarity and does not address him in the plural, which indicates a lack of respect and decency.

Moreover, 40.6% believe that in case of unfamiliarity with English, it can be replaced by another third language. On the contrary, 27.2% of respondents stated that ignorance of English completely hinders communication. Clearly in the case of Europeans traveling to English-speaking countries while not speaking English, it turns out that different relationships develop between tourists, locals, and language use. Moreover, 50.5% consider that tourists, in case they encounter a communication problem, are willing to adapt to the linguistic qualities of the destination in question. In addition, according to an experimental study conducted with visitors to New Zealand, it seems that for many tourists there is a desire to adapt to the language of their hosts. This is also argued by Cohen and Cooper (1986) since tourists occasionally adopt the language of the locals.

Finally, regarding the final choice of communication method, the question "Which method is most used when tourists and locals interact?" is reflected. Of the

202 respondents, 106 stated that they most often use verbal communication, only one non-verbal communication and the remaining 95 used both methods together. They believe, almost unanimously, that verbal communication is most effective with 93.6% (Figure 3.5).

At the same time, 84.7% consider the coexistence of the two communication methods necessary and 66.8% speculate that they do not overshadow each other. It seems that verbal communication has an advantage because it has a more specific code (Cosmopoulos, 1990). In addition, however, non-verbal messages/techniques cover a wide range of communication and induce immediacy.

Examining the correlation between the variables in a second year, it is observed that respondents with postgraduate and doctoral education have a higher average age than the other age groups, which is expected. On the other hand, although men are much less than women, in addition to searching the internet, they prefer to ask friends and relatives about planning a trip and the destination they are going to visit with 10.9%, compared to 2.9% of women. Also, 19.2% of those who do not believe that local people make friends with tourists also believe that non-verbal behavior is not exploited, compared to the corresponding percentage of those who believe in local–tourist friendship which is only 6.1%. In fact, those who believe that a local exploits non-verbal behavior also believe that they use all practices (body movements, facial expressions, and dialogue) to communicate, in contrast to those who do not believe in the exploitation of non-verbal behavior, who in turn seem to have a greater variety of opinions, and in particular a significant proportion of them believe in communication only through dialogue (34.6%). Those who remain apathetic or passively listen to a local when he or she has the floor tend to talk to him or her in addition, respondents who believe in the forced coexistence of the two (2) methods are younger than those who believe the opposite. At the

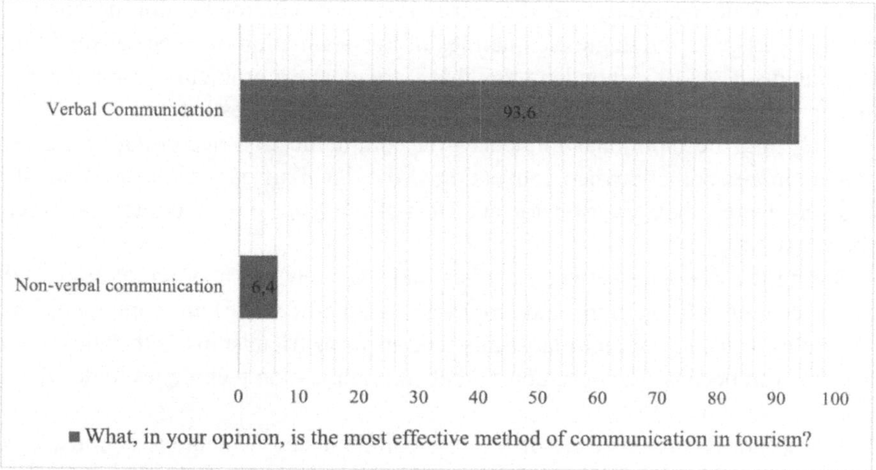

FIGURE 3.5 Bar chart of variable "Most effective method"

same time, as age increases, participation in the discussion seems to increase with interest and comments. Finally, as educational level increases, the frequency of travel increases and with this increase seems to increase the frequency of using voice tone to accompany speech. It is also logical that the frequency of using the "strong" voice tends to increase the frequency of using hand signals during speech since both practices fall within the characteristics that non-verbal communication lies in.

Conclusion

At the end of the research, the general conclusions that emerge from the analysis of the empirical part are as follows:

First of all, an admirable picture with only basic information seems to contribute more to the final choice of destination than a glowing, detailed, and highly eloquent description in words.

However, the local man uses non-verbal communication to develop a positive atmosphere between him and the tourists, and during his communication, he uses body movements and facial expressions to accompany the dialogue and also does not make friends with tourists very often, except in some cases.

At the same time, the tourist is indifferent and remains apathetic during verbal communication with locals and during the conversation with an unknown local addressing him in the singular most of the time. Sometimes this may be a sign of a lack of respect despite the fact that during the communication tourists enrich the conversation with comments and they used to speak in the plural. On the other hand, younger respondents do not participate in the conversation with such interest and comments as often as older tourists tend to do.

Regarding the spoken language, tourists demonstrate that English can be replaced by another language, as the tourist in case of communication problems is willing to adapt to the linguistic qualities of the destination or to be satisfied with body language. Most respondents say that if they have a problem communicating, they are prepared to adapt to the language qualities of the destination.

Analyzing the emotional influences on communication according to gender, men are more apathetic when conversing with a local, as opposed to women who passively watch their interlocutor and do not interrupt out of politeness except when necessary.

During the dialogue between tourists and residents, all non-verbal communication techniques (e.g. signs and emphasis on tone of voice) are combined. This appears to result in the fact that locals tend to form friendships with tourists and that's proven because they use non-verbal communication techniques to develop a positive atmosphere.

In the end, the coexistence of the two communication methods is considered necessary even though finally verbal communication appears to be the most effective method of communication.

Policy measures for more effective communication between tourists and locals

Policy measures for more effective communication between tourists and locals should have as their overriding concern the transmission of clearer and more understandable messages. Body position, hand movements, foot placement, and facial expressions can effectively contribute to this, as they convey messages that must match the speaker's words.

At the same time though a good listener should encourage the speaker to speak openly and honestly. This can be achieved by maintaining eye contact with the interlocutor, leaning slightly toward the speaker, and making gestures that show interest and attention. Frequent questions to the interlocutor can also ensure a full understanding of the message.

Regarding the communicative approach, addressing the interlocutor in the plural shows courtesy and respect for the interlocutor and makes the communication's atmosphere more appropriate.

The development of empathy on the part of the local inhabitants of a destination toward the tourist in question is a key pillar for the outcome of a more effective communication process. This is achieved by putting oneself in the other's shoes, sharing and understanding any information needs they express. Finally, those involved in communication must be jointly aware of elements of their interlocutor's background (e.g. social, ideological, cultural) so that there is respect and understanding of each other's basic ideals and that they are not trampled upon.

References

Babiniotis, G. (2022). Towards a linguistic theory of specification based on a verb grammar. *Linguistics*, *10*(3), 176–180.

Brettos, L. S. (2003). *Gamos, gennēsē, thanatos stēn archaia Ellada* (p. 109). Sabbalas.

Cohen, E., & Cooper, R. L. (1986). Language and tourism. *Annals of Tourism Research*, *13*(4), 533–563.

Cosmopoulos, A. (1990). *Relational pedagogy of the person*. Grigori.

Darwin, C. (1872). *Der Ausdruck der Gemüthsbewegung bei dem Menschen und den Thieren*. E. Schweizer.

Dilley, R. S. (1986). Tourist brochures and tourist images. *Canadian Geographer*, *30*(1), 59–65.

Floyd, J., & Fowler, Jr. (2001). *Survey research methods* (Applied Social Research Methods) (3rd ed.). Sage Publications.

Galanis, P. (2017). Fundamental principles of qualitative research in the health sciences. *Archieves of Hellenic Medicine*, *34*(6), 834–840.

Gotovos, A. (1990). *The logic of the existing school*. Gutenberg.

Heller, M., Pujolar, J., & Duchêne, A. (2014). Linguistic commodification in tourism. *Journal of Sociolinguistics*, *18*(4), 539–566.

Lickorish, L., & Jenkins, C. (2004). An introduction to tourism. *Kritiki*, *6*(2011), 3.

Littlemore, J. (2003). The communicative effectiveness of different types of communication strategy. *System*, *31*(3), 331–347.

Lytras, P. (2008). *Public relations & communication in tourism*. Interbooks.
Mill, R. C., & Morrison, A. (1992). *The tourism system: An introductory text* (2nd ed.). Prentice-Hall International Editions.
Reilly, R. T. (1988). *Travel and tourism marketing techniques*. Delmar Publishers.
Tsartas, P. (1996). *Tourists, travel, places: Sociological approaches to tourism*. Exantas, Athens.

4

IDENTIFYING WELLNESS VALUE FORMATION IN YOUNG WINE TOURISTS

A grounded theory approach

Bob Yi-Chen Duan, Jingjing Dai and Emily Ma

Introduction

"I think wine is healthier," "Wine drinking is less harmful" and "Drinking wine has so many benefits" (nine out of 14 informants; 64%). Surprisingly, drinking wine is so "**healthy**" has been overwhelmingly repeated by the informants (our data in this chapter). When comparing the literature, Chinese consumers are more concerned about the health benefits of wine than their Western counterparts (e.g. Duan et al., 2018; Liu & Murphy, 2007). The healthy consistency between the literature and the informants highlights an important research direction. However, the mechanism of such a value (the health/wellness value of wine) in the context of wine or wine tourism has not been explored in previous literature.

This chapter, thus, aims to focus on "health" from a value perspective. We define value – as a relatively stable construct that shapes individuals' beliefs, attitudes and actions (behavior) toward a specific object and contains the perception/feeling of a certain objective/activity (Hansen, 2019; Schwartz, 2021). Predominantly, tourism research has studied the formation of values. Two approaches exist as follows: service-dominant logic (SDL) or customer-dominant logic (CDL) in the service marketing domain (e.g. Grönroos, 2008).

In the context of wine tourism research, the **wellness value of wine** (WV) first received limited research attention but was rather a finding of wine tourists (e.g. Duan et al., 2019). Duan and his colleagues found that the Chinese were concerned about the WV. Secondly, in line with the value formation approaches (i.e. SDL, CDL), the logical nature of the WV has not been questioned/researched. As SDL claims, the company/provider (i.e. wine manufacturers) leads/creates the value

DOI: 10.4324/9781032633374-5

(Grönroos, 2008). Thus, this has left an essential research gap and we propose our research questions (RQs).

RQ1. Is wellness a deliberate value?

That is, is the WV purposely made by the providers for their customers? Which means providers are using wellness as a unique selling point and as an effective marketing tool to target Chinese customers in particular? The customer-to-customer (C2C) approach to value formation, on the other hand, has received more research attention in recent years (e.g. Rihova et al., 2018). C2C, as a CDL-based ontology, aims to understand the "uncontrollable" aspect (e.g. not controlled by the providers) of service marketing. In this vein, wellness value is subjective and socially constructed rather than a pure marketing tool (Nicholls & Mohsen, 2019). However, the factors that created this WV remain unanswered. Thus, a second RQ is raised.

RQ2. What factors contribute to the value formation of wellness?

This chapter has three contributions. We identified, in the first place, the underlying logic formation of the wellness value of wine (WV) and, next, the factors that influence such value (not the dimension of WV). Additionally, the proposed conceptual framework of WV is not merely applicable to the wine tourism industry to assist the decision-making process (Valeri, 2021) but also renders more efficient knowledge/information communications between different stakeholders (customers, authorities and wine industries; Valeri & Baggio, 2022).

Literature review

Value and value formation approaches

Value can be annoyingly overcomplicated. According to the theory of basic human values (Schwartz, 2021), values are defined as trans-situational goals. That is, a different situation means different values. In Schwartz's theory, the nature of value has six categories: beliefs, desirable goals, exceeding specific actions and situations, standards/criteria, organized by importance and multiple values within an action. To exemplify, values can be desirable goals to achieve (as motivation). People chase social order; that social order is the value and also motivates the action of the pursuit of it. Values also refer to standards in a different context (i.e. people judge what is good or bad) (Schwartz, 2021). Schwartz continues by claiming the importance of values varies from person to person based on a coherent structure in which the underlying foundation is either a social and psychological conflict or congruity. At the same time, some universal values are also conceptualized by Schwartz (2021), who states these universal values – such as achievement, security and power – are in line with three basic rules of human existence (i.e. individual needs of biological organisms, social interaction requirements and group survival and welfare requirements).

From a cognitive perspective, value is a trade-off between perceived gains and losses, (Aum et al., 2021). Baishya and Samalia (2020) stress that value, based on a monetary standpoint, is the comparison of the price and the product quality. Suhartanto et al. (2020), extended the monetary approach to the tourism context; that value is the destination's perceived quality with its associated price, which is realized in the visitors' evaluation of paying and receiving at a destination. In the service context, "value" means a positive feeling about oneself after the service encounter (Grönroos, 2008). The service encounter is either through self-service, such as at a buffet, or a full-service process, for example, when dining out at a restaurant. Grönroos underlined that the service or good/product is not the main interest of the customer, but rather that using the service/good is a means to reach their expected value. Despite how sophisticated value is, current tourism studies have undertaken value or perceived value as an effective variable/concept to predict customers' needs and better design the company/destination's product (Rihova et al., 2018). Such practices follow the findings from decision-making/consumer behavior studies (Ajzen, 2020). To illustrate, Rousta and Jamshidi (2020) identified five food values based on the consumption value theory, in which taste, health, price, emotion and prestige values positively influence visitors' attitudes toward selecting the food tourism destination. Rihova et al. (2018) studied the value outcomes during certain social practices (where or in what activities; festivals in their study) and concluded that effective, social, functional and network values are the result of C2C creation.

Two recent systematic literature reviews outline trends in value studies in tourism. The most common way is to examine the antecedents and/or consequences of values, such as by investigating the dimensions of personal values and their impact on attitudes and/or behavior (Kim, 2020). Carvalho and Alves (2023) confirmed Kim's systematic literature review work. Carvalho and Alves put forward that "customer behaviour" and "other factors" together created customer value based on 216 validated articles' outcomes. These "customer behaviours" are networking and customer participation, while "other factors" are personalization, emotional bonding and collaboration. Adopting a holistic view, two main approaches contributed to understanding value formation in service marketing literature, namely service-dominant logic (SDL) and customer-dominant logic (CDL). SDL is also known as "provider-dominant logic" or "goods-dominant logic." Each approach used different ontologies; SDL focused on the co-creation of value by both providers and customers (Vargo et al., 2008). Vargo's early work (Vargo & Lusch, 2004) is the start of the international debate on which service perspective can enrich service marketing. The reasoning behind SDL is that customers use resources (e.g. a table in a café), which are normally resources proposed by providers. Customers are required to possess certain skills to access/activate the resources through the service (Grönroos, 2008). The ATM in banks is an example of such logic, as customers need to understand how to use an ATM (as a skill) and then activate the "plastic card" as one resource through the self-service offered by the banks.

CDL emphasizes customers' reality and daily encounters. The foundation of such an ontology is an interpretative approach, namely that reality is socially constructed and experienced (Spivey, 1997). Such an ontology denies that reality can be known independently; it requires an interaction between a first-order reality and a second-order reality (see details in Spivey, 1997). First-order reality is reachable, such as through using languages, where people share an understanding of a thing/phenomenon (e.g. death) among each other. For the second-order reality, people in certain societies present a value/meaning/attachment to the first-order reality's phenomenon (e.g. death is an end/a new start). Briefly and easily, the world is not purely subjective but is a relative world where values are dynamically incorporated with reality. In other words, a value (how people understand death/death is sad) is potentially ever-changing via continuous destruction and construction with reality; this value is always reconstructing itself through cognitive and/or affective processes, consciously and/or unconsciously (death is an endless loop without emotional attachment).

Wellness value in wine

Wellness value in general tourism studies is centered on medical tourism, especially for international traveling behavior, such as people in rich nations seeking medical services in a less developed country (Dillette et al., 2021). Other themes include SPA/yoga/spirituality destinations (Nicolaides & Grobler, 2017), destination management and marketing (Voigt & Pforr, 2013), motivation of wellness tourism (Lim et al., 2016), conceptualization of wellness tourism (Voigt, 2014), niche market/certain market analysis (Lehto et al., 2006), mountain wellness tourism (Pan et al., 2019) and cross-cultural comparisons (e.g. Chinese and American; Han et al., 2020) and so on.

Only a few endeavors, however, exist toward the wellness value of wine tourism, while a large body of literature has precisely focused on the health benefits of wine consumption. Guilford and Pezzuto (2011) pointed out that wine drinkers tend to exercise more, are more health conscious and that regular wine consumption improves an individual's health. A psychological function was confirmed by Mortensen et al. (2001), who suggested wine drinking is closely linked with positive social, intellectual and personality benefits. More recent studies also indicate wine is associated with human health conditions such as cardiovascular health (e.g. Castaldo et al., 2019). Liu and Murphy (2007) stated that the Chinese purchase red wine for perceived health and symbolic meaning (red wine as luck). Within wine tourism studies, Frochot (2009) is the first of its kind and claims the health benefits of wine also contribute to the motivation of tourists who visit French wineries. Whereas most authors emphasized the "French paradox" in their studies, that is, the French diet contains more fat but French people have less cardiovascular disease (e.g. Renaud & Gueguen, 2007). Latter studies have not explored the health benefits of wine/WV specifically in wine tourism. In North America, the local wine industry has recognized the health value of wine, which undeniably has an important role in the industry (Getz et al., 1999). Pforr et al. 's (2016) work in the wine region in Western Australia discussed the eight local core resources that are considered vital elements for wellness destination management. Besides the above mature market (either on the supply or demand side), niche studies in Asia have a slightly different conclusion of WV. Duan et al. (2018) declared due to Chinese consumers care about the health benefits of wine, regardless of their limited wine knowledge. Thus, this value (health) has industry implications for service providers. Elsewhere in Asia, Kotur (2022) analyzed the experiences of Indian wine tourists, using Dunn's (1959) four dimensions of wellness (i.e. body, mind, spirit and environment) to identify how these four dimensions work during wine tourism experiences. Kotur concluded that the wine tourism experience stimulated the mind, body and spirit refreshment of Indian wine tourists.

To recap, two different value formation approaches (SDL and CDL) prevail in the service marketing domain, reflecting distinct worldviews. As a result, in the context of tourism, including wine tourism, the WV has not been explored regarding which worldview is leading/guiding the formation of wellness values. Further, the existing studies about WV did not specifically emphasize WV. Thus, such a research gap needs to be urgently addressed.

Method

Grounded theory

We employed traditional grounded theory (GT) as the research method. GT was developed by Glaser and Strauss (2017), as an inductive approach to generating a theory and studying a phenomenon. GT as the opposite of the dominant quantitative approach lighted another way of an unbiased method of understanding the "truth" of the world. This method requires rigor and a detailed research process (before, during and after data collection), which enables comparative analysis (Glaser et al., 2013). The comparative analysis asks for immediate data comparison within the same data, such as an emerging code, from a part of the data served to compare with the rest of the data to test/identify the differences/similarities/variations. The overarching aim of this method is to generate an abstract understanding toward a phenomenon in a systemic process. Three different genres exist in GT: traditional or classic GT considers the pattern of behavior, which creates a conceptual theory toward the behavior (Glaser et al., 2013). The second is the evolved GT (EGT), founded on the sociological perspective of symbolic interactionism (Walker & Myrick, 2006). This approach relies on the subjective meaning that people believe is true (Adeoye-Olatunde & Olenik, 2021). Actively reviewing the literature during the whole process (before, during and after data collection) is the common technique for EGT (Strauss &

Corbin, 1998). As such practice enables/enhances the theoretical sensitivity, that similar former cases/studies yield ideas/themes for following data interpretation. The third one is constructivist GT (CGT), which emphasizes people's construction of meaning concerning the inquiry area (Charmaz, 2017). CGT tends to examine a preconception and the researchers themselves are the part of their "investigation" that asks for more involvement from the researcher (Charmaz, 2017). Charmaz asserts the findings derived from the data are a construction, rather than a discovery. The research is part of the findings, as the research's background (e.g. position, experience) affects the finding formation/data interpretation. Glaser et al. (2013) argue CGT needs an agreement between informants and researchers. In line with our RQ, additional, no former literature of WV in wine tourism, particularly the value formation guided this chapter. Thus, the traditional GT was chosen, to generate the conceptual model for the WV in the context of wine tourism. The traditional GT focuses on emerging trends and ideas during the whole analytic process that addresses thinking about, constructing and interacting with the data (Morse et al., 2021).

Sampling and data analysis

Purposive sampling is applied in this chapter. This sampling enables the researcher to find the "right" people who have related knowledge to the research interests (Campbell et al., 2020). For this study, we selected two groups of informants. As we are interested in which logic (SDL or CDL) impacts value formation, both kinds of "logic-determiners" (providers vs. customers) have been added to the sampling. The first group is providers (wine manufacturers/wineries). We interviewed three managers at three different Chinese wine companies (two in Yunnan Province and one in Ningxia Province, China). To expand our understanding from the provider's side, we searched for relevant marketing information/strategies (in the Chinese market) through online information based on one French wine company using the netnography approach (Kozinets, 2015). During this online data collection process, we searched the company's website, the main shopping platforms' information and other relevant online resources (e.g. news, magazines, reports).

The second group is wine tourists (people who have visited/are visiting the Mi-Le wine region in Yunnan, China). 14 informants were selected during this process. Finally, three managers from Group 1 and 14 informants from Group 2 were included in the sample (Table 4.1). To avoid expectation bias (Adeoye-Olatunde & Olenik, 2021), the researcher told the managers that their names or company names would not be shown in publicity. The same approach was applied to informants. Further, the research explicitly told all the participants, "We are looking for real or honest answers," "we are not indicating the company's name to avoid unnecessary hassle to you and your company" and "Please feel no pressure about if your answer is right or wrong." Semi-structured questions were used for both groups (Adeoye-Olatunde & Olenik, 2021). The sample questions are shown.

TABLE 4.1 Profile of the informants

Managers	*Job task/role*	*Working experience*	*Wine tourists*	*Numbers/ frequencies*
A manager	Selling	5 years	Male, female	8, 6 (57%, 42%)
			Ages	
			18s–30	8 (57%)
			30s–40	6 (42%)
B manager	Operation	3 years	Income level monthly (RMB1)	
			Less than 1,000	2 (14%)
			1,001 to 3,000	4 (28%)
			3,001 to 6,000	4 (28%)
			6,001 to 9,000	4 (28%)
			9,001 and above	0
C manager	Operation	2 years	Days spent in the wine region	
			1 day	12 (85%)
			2–3 days	2 (14%)
			4 days and more	0

Note: 1. 1 RMB/CNY = 0.14 US dollars, 1,000 RMB =137 US dollars

For wine industries/wineries

- What kind of value are you selling to your customers?
- Do you mention the wellness/health value of wine in any marketing process?
- What do you think the "wellness value" means for the customers?

For wine tourists

- What kind of value do you think wine is associated with?
- Do you think wine has wellness benefits and why?
- Where do you obtain the wellness value of wine?
- Is the wellness value of wine important to you and why?

Data analysis was carried out in three steps: initial coding, intermediate coding and advanced coding (Tie et al., 2019). Initial coding was used at the beginning of the data collection to identify categories based on the data and guide the data collection. In Step 2, intermediate coding, the core categories were determined from the former coding and it was also determined if data saturation had been reached (Mwita, 2022). That is, after this coding process, no further data collection was undertaken at this stage. Advanced coding involves storyline and theoretical coding (conceptualizing the earlier categories into themes/patterns or theories).

Results

RQ1: Is wellness a deliberate value?

We uncover value formation through two main logics: SDL and CDL. The following section starts with SDL. Meanwhile, we divided the findings into "before visiting" the winery and "during the visit" at the winery/wine tourism experience (Table 4.2).

SDL was named PDL to specifically highlight the role of the providers rather than merely looking from the service/product point of view. Due to the pressure of (potential) punishments from the local rules and regulations, or in a safer way (in this data site; China), the providers did not promote any health benefits of wine consumption. Manager A (male, 26-year-old) commented, "We are following the rules . . . we did not post anything associating wine with health." Manager B (female, 29-year-old) added, "Some Chinese Bai-Jiu (strong liquor) you see on the TV, they are saying their Bai-Jiu boosts your health and has this and that benefit." According to local advertising laws, alcohol companies cannot conduct any advertisements claiming that alcohol is good for one's physical body. However, based on the data, some players/providers ignore such laws, or play in a "smart way." Manager B continued, "I am not sure, but from my understanding, these Chinese Bai-Jiu

TABLE 4.2 Wellness value logic

Before Visit			
Provider-dominant logic (PDL/SDL)		*Customer-dominant logic (CDL)*	
Foreign brand	*Chinese brands*	*Pre-existing wellness value*	*Other values*
• History • Occasion • Wine information (e.g. Grape types, aroma). • Wine region information • *Pair with Chinese and Western foods*	• History • Occasion • Wine information (e.g. Grape types, aroma). • Wine region information • Pair with foreign foods • Foreign winemakers	• Family background (frequency/f, 67) • Social interactions (f, 89) • Internet (f,24)	• Packaging (f, 182) • Taste (f, 61) • Aroma (f, 70) • Smooth (f, 63) • Strong flavor (f,18) • Clear color (f,21) • Fruity (f, 170)
During Visit			
PDL in wine trips Educating wellness value (Interactions between the company and the customers)		CDL in wine trips Reinforcing wellness value (Interactions among the customers)	

companies did not say the health benefits directly . . . they normally mention our Bai-Jiu nourishes men's kidneys." Such circumvention of the banned behavior was indicated by the wine providers, who indicated that they were not planning to do so or did this (marketing the health benefits of wine in public channels).

In contrast, Chinese Bai-Jiu companies get around the ban by not straightforwardly, saying their alcohol is good for health but emphasizing the kidney. This may be due to the value of traditional Chinese medicine, which holds that the kidney is linked with reproduction ability. Simultaneously, most Bai-Jiu customers were Chinese males (Qian et al., 2015). These males were inclined to interpret the information that Bai-Jiu was equal to kidney issues, which benefits their reproduction ability. On the other hand, wine companies did not stress such "kidney issues," maybe because their target customers are different from Bai-Jiu companies.

One interesting finding is that the Chinese and French wine companies promote slightly different values/information to their customers. The Chinese wine companies presented how to pair their wines with Western foods and also stressed that their companies hired foreign winemakers to guide their quality. In contrast to Chinese companies, the French wine company paired its wines with Chinese foods. Four factors/values are mentioned by all the wine companies: history, the occasion of consumption, wine information (e.g. grape type, aroma) and wine region information. This finding indicates that "wine and its related information" are considered important from the wine providers' perspective. In line with SDL (Grönroos, 2008), the service providers offer a resource that the customer could use to generate value. In other words, before the wine tourism experience, the providers did not have any resources (marketing sources) that supported wellness value formation, or at least on the surface, there is no clear evidence they did so.

During the wine tourism experience, the co-creator role of providers appears. Even though SDL (Grönroos, 2008) is provider-leading logic, this logic also indicates that providers tend to be the value co-creators; the customers are the foundation that produces values. Education about wine-related knowledge for tourists is available at wineries. This reflects early work showing that wine tourists have a desire to learn new things (Duan et al., 2019). Importantly, wellness value is also included in this experience, in which the "interaction" between providers and tourists is the process for the co-creation of value, as the winery has tour guides within the winery. Manager C (male, 34-year-old) talked about how "our tour guides will include the wellness value during their speech." Inside the cellar door, the winery has posts and boards showing wine-related information and wellness value are one kind of that information. Hence, active interaction (with the tour guide) and passive value co-creation (posts and boards) co-exist.

Wellness value from the CDL perspective (Heinonen et al., 2010) is focused on the customers. The informants have wellness values in their daily practices. Thus, this finding also confirms that wellness value is not simply SDL-based (i.e. provider-led value). However, if this research stops at "before winery visit," the conclusion of wellness value is CDL. That is, if we only consider value creation before the winery visit, wellness value can be seen to follow a customer-dominant logic. Three factors led to value formation: family background, social interaction and learning from the Internet. Furthermore, during the winery visit, the informants commented that they had deepened their understanding of the wellness value of wine. This is because of the aforementioned educational experience offered by the winery. Another factor is that interaction among the wine tourists themselves also contributed to/enhanced the wellness value of wine.

RQ2: What factors contribute to the value formation of wellness?

This RQ focused on the dimensions contributing to the formation of wellness value rather than the dimensions that make up the construct of wellness. Based on the data, three factors/dimensions influenced wellness value formation: **wine, reality and life** and **intangible culture**, as in Table 4.3.

Wine includes "wine-drinking behavior" and "curiousness" as its sub-categories. For example, during social get-togethers, wine tourists encounter wellness value creation. Wine tourists can either be givers or recipients of value. Givers normally introduce the wellness value. For example, as one informant (female, 24-year-old) mentioned, "My uncle talked about the wellness value of wine. . . . I was not sure about this wellness thing before. . . . Later, I also talked about these things with others." Here, this informant is both a giver and a recipient.

Reality and life contain two categories: "work-related" and "private social life-related." This factor is largely reflected in the CDL in that this logic is based on the reality and daily lives of the customers. This theme indicated that the informants' daily lives involved a lot of social interaction, especially work-related dinners. One informant (male, 30-year-old) said,

> I seldom eat at home and normally I have dinner at home maybe once a month. . . . Of course, we drink a lot of alcohol, but wine is mainly for women to drink." "Wine is almost like a soft drink and good for their health.

TABLE 4.3 Factors contributing to wellness value

Themes	*Categories*	*Sub-categories*
Wine	Wine-related	• Wine-drinking behavior • Curiousness
Reality and life	Work-related	• Frequent work banquet • Drinking habit/culture
	Private social life-related	• Frequent social gatherings (friends, relatives) • Peer influence
Intangible culture	Culture-related	• Traditional Chinese Medicine value • De-stress • "Soft drink"/not harmful

Intangible culture has one category: "culture-related." Within this theme, the informants mentioned that moderate wine drinking is a good behavior that keeps their bodies healthy. One informant (female, 23-year-old) explained, "I think maybe that traditional Chinese medicine (TCM) value impacts the wellness value of wine." "You see, TCM encourages people to drink a bit of alcohol to maintain health." Similarly, for Western wine drinkers, wine helps to relax the informants and is another sub-category within this theme (Duan et al., 2018). Additionally, half of the male informants (4, 28%) think wine is a "soft drink" and even state that alcohol is "water." This is a unique finding that brings a new understanding of alcohol consumption associated with its perceived value/value.

Discussion

This chapter aims to identify the underlying logic of the wellness value of wine (WV) formation, especially given the profits/revenue that wine companies can gain through actively marketing WV. As suggested by Duan et al. (2018), the health value of wine among Chinese wine tourists has an industry implication for service providers. Meanwhile, the American wine industry agreed that the health value of wine plays an important role in the industry (Getz et al., 1999). In the existing literature, however, the mechanism of WV formation, particularly in the context of wine tourism, is absent. Furthermore, two dominant approaches to value formation exist in tourism studies: SDL and CDL. In the meantime, factors that determine value formation are largely ignored. In other words, the question of which logic (SDL or CDL) determines the WV has not been answered. Thus, two RQs guided this chapter. In RQ1, we answered the formation logic of WV. In the second RQ, we identified the factors contributing to the WV. Consequently, two theoretical contributions are made. Besides, the methodology adopted online and offline data and enriched the findings and literature. Furthermore, integrating the results of two RQs, we propose Figure 4.1. This proposed framework demonstrates the logic of WV and may also apply to understanding the general Chinese public regarding their value formation in a general tourism context.

This proposed framework has at least three contributions. Firstly, a new construct has been added to the proposed model. Traditional CDL emphasizes that the value of customers comes from "reality and living" (customers' daily lives), while SDL puts the research effort into the resources offered by the providers (Grönroos, 2008). We added culture to the whole value formation process, and our findings also indicate that SDL and CDL in certain contexts (e.g. wine tourism), cannot be treated independently. A combination of three (i.e. CDL, SDL, and culture) extends our understanding of value formation. In other words, the proposed model highlights the complexity of value formation that a single approach (either CDL or SDL) does not analyze deeply (e.g. WV formation). Further, single-approach-based

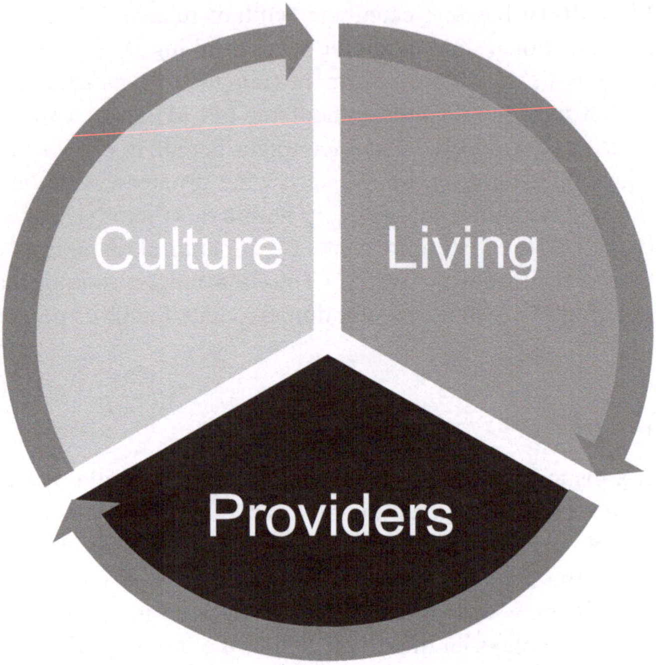

FIGURE 4.1 Conceptual framework of wellness value formation in wine tourism

research in this context will not consider value formation from a holistic point of view. Subsequently, further studies are encouraged to apply a combination of CDL and SDL to investigate value formation as well as add necessary extras (i.e. culture in this chapter).

Secondly, the relationship is dynamic and multi-dimensional. The Chinese young wine tourists (18-year-old to 39-year-old) have WV, but their level of understanding and concern toward WV varies. The male informants tend to think wine is a "soft drink" or like a bottle of "water." Thus, their "living" dimension in Figure 4.1 has a weak influence on their WV. This means wine in their daily practice has not determined its wellness value. However, Chinese Bai-Jiu (strong liquor) providers promote "kidney" issues that are related to reproduction ability and seem to have more "wellness" value formation. Even though Bai-Jiu companies should not get around the ban on local rules and regulations, the "kidney" issue is indeed a successful marketing strategy. Also, Figure 4.1 contains multiple dimensions in its model (as mentioned in the earlier paragraphs). The model has three dimensions, and we further address the importance of the "living" dimension. In the context of wine tourism, the "living" dimension dominated the WV, as found in RQ1. Meanwhile, two other dimensions (culture and provider) facilitated the completed value formation. The following studies are encouraged to apply Figure 4.1 to wine tourism destination management with social network analysis (Valeri & Baggio, 2020) to test/examine WV or other values of wine that can be used by the industry. Besides, based on the multi-dimensional nature of Figure 4.1, promoting the vineyard culture/wine-related culture/wine region culture seems like an effective marketing strategy from the providers' sphere (Valeri & Fadlon, 2016).

Thirdly, we confirmed that CDL is the dominant approach to value formation. The customers are the leaders of value formation in the tourism context. These findings support earlier studies (e.g. Heinonen et al., 2010). We contend CDL is still a promising value formation in the service marketing domain. However, as pointed out in previous contributions, a pure CDL is insufficient to understand value formation because the complexity, dynamics and uncertainty of the real world cannot be simply understood through a sole approach.

Conclusion

This chapter uncovers that the wellness value of wine (WV) is not deliberate (RQ1). By deliberated value, we mean a value created mainly from the provider's side (SDL). This wellness value is largely based on individuals themselves: the individuals' daily lives, such as their work type, personality and social circles (in our categorization, "reality and life"). The wine companies and wineries, as the actors in the other logic formation approach (SDL), are confirmed as value co-creators; this finding is consistent with early studies on value formation in the tourism context (e.g. Grönroos, 2008). Furthermore, during the wine tourism experience, providers interact more with their customers and deepen the wellness value. RQ2 identified that "Wine," "Reality and life" and "Intangible culture" are the factors that contributed to wellness value formation. In summary, as a result, these two RQs present two theoretical contributions. The proposed framework makes three more contributions. Last but not least, online and offline methods offer a richer picture of the providers' sphere. This chapter contains two main limitations. First, the pre-existing CDL and SDL approaches to identifying value formation may be imperfect/have a bias in understanding complex value formation from an open-minded perspective. An exploratory approach is advised for further study. Moreover, sample size, location and diversity are not necessarily large enough and may have potential bias. Thus, sampling improvement is the next recommendation.

Reference

Adeoye-Olatunde, O. A., & Olenik, N. L. (2021). Research and scholarly methods: Semi-structured interviews. *Journal of the American College of Clinical Pharmacy*, *4*(10), 1358–1367.

Ajzen, I. (2020). The theory of planned behavior: Frequently asked questions. *Human Behavior and Emerging Technologies*, *2*(4), 314–324.

Aum, S., Lee, S. Y. T., & Shin, Y. (2021). Inequality of fear and self-quarantine: Is there a trade-off between GDP and public health? *Journal of Public Economics*, *194*, 104354.

Baishya, K., & Samalia, H. V. (2020). Extending unified theory of acceptance and use of technology with perceived monetary value for smartphone adoption at the bottom of the pyramid. *International Journal of Information Management*, *51*, 102036.

Campbell, S., Greenwood, M., Prior, S., Shearer, T., Walkem, K., Young, S., . . . Walker, K. (2020). Purposive sampling: Complex or simple? Research case examples. *Journal of Research in Nursing*, *25*(8), 652–661.

Carvalho, P., & Alves, H. (2023). Customer value co-creation in the hospitality and tourism industry: A systematic literature review. *International Journal of Contemporary Hospitality Management*, *35*(1), 250–273.

Castaldo, L., Narváez, A., Izzo, L., Graziani, G., Gaspari, A., Di Minno, G., & Ritieni, A. (2019). Red wine consumption and cardiovascular health. *Molecules*, *24*(19), 3626.

Charmaz, K. (2017). The power of constructivist grounded theory for critical inquiry. *Qualitative Inquiry*, *23*(1), 34–45.

Dillette, A. K., Douglas, A. C., & Andrzejewski, C. (2021). Dimensions of holistic wellness as a result of international wellness tourism experiences. *Current Issues in Tourism*, *24*(6), 794–810.

Duan, B., Arcodia, C., & Ma, E. (2019). Positioning the current development of China's wine tourism destinations: A netnography approach. In *Wine tourism destination management and marketing* (pp. 537–550). Palgrave Macmillan.

Duan, B., Arcodia, C., Ma, E., & Hsiao, A. (2018). Understanding wine tourism in China using an integrated product-level and experience economy framework. *Asia Pacific Journal of Tourism Research*, *23*(10), 949–960.

Dunn, H. L. (1959). High-level wellness for man and society. *American Journal of Public Health and the Nations Health*, *49*(6), 786–792.

Frochot, I. (2009). Wine tourism in France: A paradox? In *Wine tourism around the world* (pp. 67–80). Routledge

Getz, D., Dowling, R., Carlsen, J., & Anderson, D. (1999). Critical success factors for wine tourism. *International Journal of Wine Marketing*, *11*(3), 20–43.

Glaser, B., Bailyn, L., Fernandez, W., Holton, J. A., & Levina, N. (2013). What grounded theory is . . . In *Academy of management proceedings* (Vol. 2013, No. 1, p. 11290). Academy of Management.

Glaser, B., & Strauss, A. (2017). *Discovery of grounded theory: Strategies for qualitative research*. Routledge.

Grönroos, C. (2008). Service logic revisited: Who creates value? And who co-creates? *European Business Review*, *20*(4), 298–314.

Guilford, J. M., & Pezzuto, J. M. (2011). Wine and health: A review. *American Journal of Enology and Viticulture*, *62*(4), 471–486.

Han, H., Kiatkawsin, K., Koo, B., & Kim, W. (2020). Thai wellness tourism and quality: Comparison between Chinese and American visitors' behaviors. *Asia Pacific Journal of Tourism Research*, *25*(4), 424–440.

Hansen, A. V. (2019). Value co-creation in service marketing: A critical (re)view. *International Journal of Innovation Studies*, *3*(4), 73–83.

Heinonen, K., Strandvik, T., Mickelsson, K. J., Edvardsson, B., Sundström, E., & Andersson, P. (2010). A customer-dominant logic of service. *Journal of Service Management*, *21*(4), 531–548.

Kim, M. (2020). A systematic literature review of the personal value orientation construct in hospitality and tourism literature. *International Journal of Hospitality Management*, *89*, 102572.

Kotur, A. S. (2022). Exploring the wellness dimensions of wine tourism experiences: A netnographic approach. *International Journal of Wine Business Research*, *34*(4), 608–626.

Kozinets, R. (2015). Management netnography: Axiological and methodological developments in online cultural business research. In *The Sage handbook of qualitative business and management research methods*. Sage.

Lehto, X. Y., Brown, S., Chen, Y., & Morrison, A. M. (2006). Yoga tourism as a niche within the wellness tourism market. *Tourism Recreation Research*, *31*(1), 25–35.

Lim, Y. J., Kim, H. K., & Lee, T. J. (2016). Visitor motivational factors and level of satisfaction in wellness tourism: Comparison between first-time visitors and repeat visitors. *Asia Pacific Journal of Tourism Research*, *21*(2), 137–156.

Liu, F., & Murphy, J. (2007). A qualitative study of Chinese wine consumption and purchasing: Implications for Australian wines. *International Journal of Wine Business Research*, *19*(2), 98–113.

Morse, J. M., Bowers, B. J., Charmaz, K., Clarke, A. E., Corbin, J., Porr, C. J., & Stern, P. N. (2021). *Developing grounded theory: The second generation revisited*. Routledge.

Mortensen, E. L., Jensen, H. H., Sanders, S. A., & Reinisch, J. M. (2001). Better psychological functioning and higher social status may largely explain the apparent health benefits of wine: A study of wine and beer drinking in young Danish adults. *Archives of Internal Medicine*, *161*(15), 1844–1848.

Mwita, K. (2022). Factors influencing data saturation in qualitative studies. *International Journal of Research in Business and Social Science (2147–4478)*, *11*(4), 414–420.

Nicholls, R., & Mohsen, M. G. (2019). Managing customer-to-customer interaction (CCI) – insights from the frontline. *Journal of Services Marketing*, *33*(7), 798–814.

Nicolaides, A., & Grobler, A. (2017). Spirituality, wellness tourism and quality of life. *African Journal of Hospitality, Tourism and Leisure*, *6*(1), 1–37.

Pan, X., Yang, Z., Han, F., Lu, Y., & Liu, Q. (2019). Evaluating potential areas for mountain wellness tourism: A case study of Ili, Xinjiang province. *Sustainability*, *11*(20), 5668.

Pforr, C., Voigt, C., & Locher, C. (2016). The Margaret river wine region: Analysing key resources as a basis for a competitive wellness tourism destination. In *Competence-based innovation in hospitality and tourism* (pp. 247–270). Routledge.

Qian, L., Newman, I. M., Xiong, W., & Feng, Y. (2015). Traditional grain alcohol (bai jiu, 白酒) production and use in rural central China: Implications for public health. *BMC Public Health*, *15*(1), 1–11.

Renaud, S., & Gueguen, R. (2007, September). The French paradox and wine drinking. In *Novartis foundation symposium 216-alcohol and cardiovascular diseases: Alcohol and cardiovascular diseases: Novartis foundation symposium 216* (pp. 208–222). John Wiley & Sons, Ltd.

Rihova, I., Buhalis, D., Gouthro, M. B., & Moital, M. (2018). Customer-to-customer co-creation practices in tourism: Lessons from customer-dominant logic. *Tourism Management*, *67*, 362–375.

Rousta, A., & Jamshidi, D. (2020). Food tourism value: Investigating the factors that influence tourists to revisit. *Journal of Vacation Marketing*, *26*(1), 73–95.

Schwartz, S. H. (2021). A repository of Schwartz value scales with instructions and an introduction. *Online Readings in Psychology and Culture*, *2*(2), 9.

Spivey, N. N. (1997). Reading, writing and the making of meaning. In *The constructivist metaphor*. Academic Press. doi: 10.358470

Strauss, A., & Corbin, J. (1998). *Basics of qualitative research techniques*. Sage.

Suhartanto, D., Brien, A., Primiana, I., Wibisono, N., & Triyuni, N. N. (2020). Tourist loyalty in creative tourism: The role of experience quality, value, satisfaction, and motivation. *Current Issues in Tourism*, *23*(7), 867–879.

Tie, Y. C., Birks, M., & Francis, K. (2019). Grounded theory research: A design framework for novice researchers. *SAGE Open Medicine*, *7*, 2050312118822927.

Valeri, M. (2021). *Organizational studies: Implications for the strategic management*. Springer Nature.

Valeri, M., & Baggio, R. (2020). Social network analysis: Organizational implications in tourism management. *International Journal of Organizational Analysis*, *29*(2), 342–353.

Valeri, M., & Baggio, R. (2022). Increasing the efficiency of knowledge transfer in an Italian tourism system: A network approach. *Current Issues in Tourism*, *25*(13), 2127–2142.

Valeri, M., & Fadlon, L. (2016). Sustainability in tourism: An originality and hospitality business in Italy. *Tourismos*, *11*(1), 1–18.

Vargo, S. L., & Lusch, R. F. (2004). The four service marketing myths: Remnants of a goods-based, manufacturing model. *Journal of Service Research*, *6*(4), 324–335.

Vargo, S. L., Maglio, P. P., & Akaka, M. A. (2008). On value and value co-creation: A service systems and service logic perspective. *European Management Journal*, *26*(3), 145–152.

Voigt, C. (2014). Towards a conceptualization of wellness tourism. *Wellness Tourism: A Destination Perspective*, 19–44.

Voigt, C., & Pforr, C. (2013). *Wellness tourism: A destination perspective*. Routledge.

Walker, D., & Myrick, F. (2006). Grounded theory: An exploration of process and procedure. *Qualitative Health Research*, *16*(4), 547–559.

5

DIGITAL MARKETING AND CONSUMER BEHAVIOR IN TOURISM

Tolga Gök and Mahmut Baltacı

Introduction

In the early twentieth century, when the need for digital marketing, digital promotion, digital advertising, digital distribution, and digital customer experience emerged, following the steps of technological evolution, there was a great increase in value (Sismeiro & Bucklin, 2004). The development of new technologies, and especially the internet, has encouraged industries such as tourism to adapt their business models to new forms of marketing ecosystems (Palos-Sanchez et al., 2019) in the twenty-first century (Bennett et al., 2017). Digital technology provides new communication and distribution channels for both consumers and suppliers of travel services. The internet enables travel agents to adopt digital marketing to attract, inform and serve travelers.

As of the 1960s, the discovery of the internet and the development of web technologies in the ongoing process made changes in many areas compulsory. As a result, the developments in technology, artificial intelligence, and smart systems are mentioned at the point reached today and their effects on business and life are discussed. Tourism is one of the most affected sectors by technology. The widespread use of technological devices (computers, phones, tablets, etc.) has made it necessary to switch to a technology-based marketing approach in terms of marketing tourist products. At this point, businesses aim to be one step ahead of the competition and reach wider audiences by increasing their digital marketing efforts.

It has been seen in recent years that tourism enterprises have realized the advantages of switching from traditional marketing to digital marketing and are trying to use it in the most efficient way, such as transportation, accommodation, food and beverage, entertainment, and guidance. In the planning phase of various activities, technology is used effectively to communicate with each other and with potential

DOI: 10.4324/9781032633374-6

tourists. Thanks to digital marketing, tourism businesses gain advantages in many areas such as effective advertising and promotion activities, acceleration of reservation processes, and monitoring of customer behavior. Thanks to the marketing activities carried out in the digital environment, businesses also have the opportunity to reach the cost, time and larger markets.

By transferring their activities to digital platforms, tourism businesses can convince their customers more easily and understand their expectations better. Thus, they can reach an advantageous position in ensuring customer satisfaction and loyalty. One of the most important issues for businesses to carry their marketing activities to the digital environment is to have trained and experienced staff. At this point, it is very important for businesses to invest in human resources while investing in technology on the one hand. That's why tourism businesses use digital marketing tools like search engine optimization, search engine ads, social media platforms, e-mail, etc. It is very important for them to be aware that they can convert potential customers into actual customers and increase their market share by using digital resources. Businesses that do not have this awareness are likely to lose their competitive advantage.

In terms of businesses serving in the tourism sector, the conscious use of digital marketing in communication with consumers is an important power. With the continuous development of digital marketing, studies in this field are also increasing. In this section, the importance and usage areas of digital marketing in terms of the tourism sector are emphasized.

Digital marketing: theoretical and conceptual framework

Definition of digital marketing

Digital marketing is the use of digital technologies that enter our lives with the internet in studies carried out to achieve business goals, advertising and promotion activities, image and brand management, and effective communication with consumers, especially with the target audience, as in traditional marketing (Demirdögmez, 2021). As companies in the developing world realize the importance of digital marketing, it has become necessary for them to combine online with traditional methods in order to more precisely meet the needs of their customers in order for them to be successful. Online advertising is a powerful marketing tool for building brands and for companies to succeed. Digital marketing is more cost-effective in terms of generating results and measuring the success of advertising money spent and the ROI of advertising (Bala & Verma, 2018).

Digital marketing is the use of digital technologies to create channels to reach potential buyers, to achieve business goals by meeting consumer needs more effectively. It is often considered synonymous with internet marketing or e-marketing (Sawicki, 2016). It is also referred to as "online marketing", "internet marketing" or "web marketing" in the literature. Although the term digital marketing has gained popularity over time, it seems that the terms online marketing in the USA and web marketing in Italy are more widely used. After 2013, the use of digital marketing has come to the fore as the most common term, especially in the UK and worldwide (Kaur & Sandhu, 2017).

Digital marketing differs from traditional marketing methods with its features such as providing an environment that can reach large masses with low costs and changing marketing strategies from start to finish. While promoting the brand and the work performed, it includes marketing activities carried out using the internet, interactive environment, and mobile opportunities in digital and social media in order to support all marketing activities (Chaffey & Smith, 2013, p. 102).

The term digital marketing has evolved over time from a specific term that describes the marketing of products and services using digital channels to an umbrella term that describes the process of using digital technologies to acquire customers and create customer preferences, promote brands, retain customers, and increase sales (Kannan & Li, 2017). The fact that businesses have to make changes in their marketing activities due to reasons such as speed, electronics, change, innovation, and digitalization has made it necessary to adapt to digital marketing practices. This situation increased the sales of the companies and ensured the company image and brand awareness (Demirdögmez, 2021).

Digitalization and the digital marketing methods

While digitalization gives consumers the opportunity to express themselves and voice their opinions, it also gives them the power of choice and influence. Brands have the chance to interact with customers dynamically, while users have the chance to search for information, products or services. Digital marketing gives businesses the ability to expand their reach and creates value for consumers and businesses. The primary goal of brands is to build trust and intimacy among their customers who purchase their products or services. Thanks to its personalized use, those working in the digital marketing industry can customize the content to be more human-oriented and user-friendly. Users can post comments and post content for a brand in a way that can influence people and opinions. Content, time, comments, interface, reviews, speed, and image play a key role in the digital age. Users tend to respond positively to brands when they manage to create a welcoming and customer-focused environment. User profiles, audiences, and consumer behavior engage the internet user with work and generate high-quality services, high-volume revenues, and excellent user experience satisfaction (Carvalhosa et al., 2017; Wierenga, 2010).

Digitalization has drastically changed the way companies approach the consumer. Thanks to the Internet, a consumer can analyze and sometimes even dictate a price based on the circulating opinions. The development of digital technologies has also led to the emergence of the cyber consumer and cyber business world. Digitization makes it possible to implement new relationship models with consumers, which are challenging for businesses that want to increase their market share. Smartphones and social media have had a certain impact on the dynamic development of digital marketing. Therefore, the ways to reach potential customers through mobile phones and social networks are current communication channels that can establish relationships with customers and therefore increase sales (Sawicki, 2016).

Digital marketing methods such as search engine optimization (SEO), search engine marketing (SEM), content marketing, influencer marketing, content automation, campaign marketing, data-driven marketing, e-commerce marketing, social media marketing, social media optimization, e-mail direct marketing, display advertising, e-books, and optical discs and games have become more common in emerging technology. Digital marketing now extends to non-Internet channels that provide digital media such as mobile phones (SMS and MMS), call-back, and on-hold mobile rings (Desai, 2019), and some of the digital marketing techniques are as follows (Bala & Verma, 2018).

Search engine optimization (SEO) adjusts the website to appear naturally or organically for search results on Google, Yahoo Bing or any other search engine. Google regularly updates its algorithms so that only relevant results come up. From this perspective, many experts say that SEO is dead and the effort is in vain. But the truth is that Google tries to prevent algorithm manipulation and filters out sites that do not deserve to be at the top of SERPs (Search Engine Results Pages). Therefore, it is important to invest in SEO studies. It covers technical issues related to website, content and query matching, spidering, indexing, and interpreting non-text content. Search engine optimization is the most cost-effective marketing strategy that will bring organic traffic to businesses.

Search engine marketing, or SEM, is a comprehensive strategy to drive traffic to your business primarily through paid efforts. Therefore, it is also called paid search marketing. The universe of SEM is diverse and complex. Depending on your business structure, you can choose the PPC (pay-per-click) or CPC (cost-per-click) model or the CPM (cost-per-thousand impressions) model. There are different platforms for SEM. By far, Google Ad Words (on the Google Network) and Bing Ads (on the Yahoo Bing Network) are the most popular. SEM also includes display advertising, search retargeting and site remarketing, mobile marketing, and paid social advertising.

According to Pulizzi (2007), content marketing is a marketing method that produces remarkable and consistent content in order to win customers, create loyalty, and encourage customers to buy. Content marketing refers to the creation and promotion of content assets with the aim of generating brand awareness, traffic growth, lead generation, and customers. Channels that can play a role in your content marketing strategy include blog posts, eBooks and whitepapers, infographics, online brochures, and lookbooks (Desai, 2019).

Social media marketing, or SMM, is a branch of SEM efforts. It involves managing the traffic of sites through social sites such as Facebook, Instagram, Twitter, Pinterest, Google+, and LinkedIn. Good content is shared and loved. It is important to create and customize content for different social media platforms. Generative and originality are important. It is important to interact with users at least four to five times a day, on a daily basis. Your SMM efforts can be especially useful for branding and selling. The tools and approaches to communicating with customers have changed drastically with the advent of social media. Therefore, businesses must learn how to use social media consistently with their business plans (Mangold & Faulds, 2009). To create a successful marketing campaign through social media, a consumer must be open to technology.

E-mail marketing is the strategy of sending a commercial message via email to a list of potential customers. With effective email marketing software, you can maintain email lists that are broken down by various factors, including customers' likes and dislikes and spending habits. Sending personalized emails is important, and this move helps build trust. However, it is important to note that e-mail marketing can also be considered spam and in some countries, there are laws against it.

Consumer behavior in digital tourism

Digital technology is used in marketing to make tourism offerings competitive and meet the expectations of digital consumers (Happ & Ivancso-Horvath, 2018; Sharma et al., 2020). Communicating to customers about products and services through digital media and making the interaction between suppliers and customers seamless is the main goal of digital marketing (Yasmin et al., 2015). Digital marketing can be seen as activities, institutions, and processes facilitated by digital technologies to create, communicate and deliver value for customers and other stakeholders.

Digital technology has an important place in terms of the globalization of tourism. Digital technology is providing new changes in marketing strategy. Several opportunities and threats to the tourism industry occur (Gratzer et al., 2004). The tourism industry is highly open to the benefits of digital technologies due to their fragmented and knowledge-driven nature (Buhalis & Deimezi, 2004). Travel service providers use digital tools such as websites, mobile apps, and social media to attract new consumers (Mariani et al., 2014), improve customer service, improve operational efficiency, and ultimately increase revenue (Amaro & Duarte, 2015; Law et al., 2014).

Digital tourism is the use of digital technologies to increase and improve the experience gained by tourists. It also blends the real world with digital content to enhance the tourist experience. Mobile phones and the internet have been driving the digital travel revolution in many developed countries in recent years, bringing the tourism industry more closely together with digital tourism (Law et al., 2018; Reino et al., 2013). Mobile phones and the internet enable tourism destinations to increase the online presence and offline connectivity necessary to compete in today's global tourism market. They provide the necessary tools and practices to handle tourism management more efficiently; offer cheap access to existing and potential customers; and enable multiple distribution channels (Buhalis, 2003; Law et al., 2018). Smartphones have become an integral part of everyday life, and a lot of people, including tourists, use the latest technology. With the increase in having multiple technological devices and the contribution of mobile travel bookings to the overall travel market, smartphones and smart tourism have now become powerful tools for tourists thanks to mobile applications that make it easier for them to travel (Law et al., 2018; Murphy et al., 2016). Travel-based mobile apps are one of the most downloaded category apps, and an increasing number of smartphone users are opting for travel apps to plan their leisure tours. Tourism mobile apps help users with trip planning, route mapping, ticket reservations, accommodation reservations, taxi bookings, and more (Dickinson et al., 2014; Wang & Fesenmaier, 2013; Wang et al., 2012). The tourism industry uses apps to get involved in the market and reach the next level of user interaction.

The travel industry plays an important role in the formation of the tourism movement. The current use of Travel 2.0 and the development of Travel 3.0 will be important in advising tourists' travel experiences and guiding tourists in future. Tourists can make their travel plans and comment on touristic products with the help of the internet. In addition, the photos they upload on social media will guide other tourists. With the help of Travel 3.0, the most suitable and according to their needs travel plans will be presented to the tourists (Laboy & Torchio, 2007).

In tourism, both social media and word-of-mouth marketing are important in purchasing tourist products. The importance of the internet in the use of social media is very great. Especially, wireless internet has become a must for tourists. According to the 2021 Wi-Fi-Alliance annual report, 16 billion Wi-Fi devices were used in 2021. It is thought that Wi-Fi technology will provide an input of 5 trillion dollars to the world economy by 2025 (Wi-Fi Alliance, 2022).

Altınay et al. (2017), in their study, aimed to reveal the effect of social media on tourists' touristic product purchases and to reveal the changes in consumer movements. They also aimed to find out how often and what tools social media is used. At the same time, it is aimed to find the role of social media before and after the purchase of the touristic product. Data were collected from 455 people by the authors using the questionnaire technique. Facebook has been the most and most frequently used platform. It has been found that social media users are influenced by social media comments and they follow social media comments while buying tourism products.

Chen and Tsai (2008), in their study, aimed to examine the relationship between perceived value, customer satisfaction and customer loyalty in social media tools in travel marketing. They reached 407 people with the questionnaire technique. It has been found that the value in the travel market comes to the fore in TV marketing, the perceived risk is high in TV shopping due to the intangibility of the travel product, but the satisfaction and loyalty from TV shopping are high.

Laajini (2021), in his study, focused on Moroccan tourists' use of new technologies and online touristic product purchasing experiences. It has been tried to measure the interaction, use, and trust of tourists with digital platforms. Data were obtained by asking questions to four focus groups between the ages of 16–64. Online providers have been found to be an integral factor in the behavior of Moroccan tourists as they facilitate research and save time.

Kim et al. (2008), in their study aimed to examine tourists' mobile device usage and tour purchasing experiences. In the study, the extent to which tourists accept and use mobile devices was examined. The survey was applied to people who bought tours from e-rewards.com and 283 people received full answers. The results show that travelers' technology use and travel experience are effective factors in mobile technology adaptation. In addition, it has been found that frequent travelers have more positive attitudes toward using mobile devices and have a higher intention to use mobile devices while traveling.

Studies show that the use of social media has an important place in tourism marketing. It is also seen that travelers use mobile devices more frequently during travel and their purchasing experiences are based on comments and opinions on social media.

Conclusion

Tourism businesses struggle in an intensely competitive environment to achieve their goals. Although these purposes are basically the making of money and the continuity of the enterprises, they can be expressed as meeting the expectations in the service sector and ensuring customer satisfaction. In today's world, consumers are more conscious and it is not easy to be persuaded. It is very important for the customer to have the products they demand quickly and without any problems. Thanks to technology, the question marks about the products in the minds of the customers are also relatively reduced. The products offered in the service sector are intangible by nature, and there is a benefit obtained when purchased. Considering that the quality of the products offered in the tourism sector is similar, it is a necessity to benefit from technology. Laying the foundations of the image that will be formed in the minds of customers begins with the use of technology.

Digital marketing, which has become important in communicating with customers, will ensure the continuity and loyalty of the customer. Businesses that do not have an innovative perspective will struggle with traditional marketing methods, while technology-sensitive competitors will go a long way. When digital marketing activities for the promotion of tourist destinations are supported with virtual and augmented reality-like technologies, their effect will increase significantly. Tourism enterprises should be encouraged to understand the importance of digital marketing activities in increasing their market share and making the necessary investments. NGOs in various fields of tourism should take an active role in ensuring this. Opening training programs on digital marketing and presenting them in educational institutions at various levels will support meeting the need for human resources in this field. As in every field, it is foreseen that the existence of tourism businesses in the digital world today and in future can be achieved with different thinking, having an innovative perspective, and a visionary mindset.

References

Altınay, M., Güçer, E., & Bağ, C. (2017). Consumer behavior in the process of purchasing tourism product in social media. *İşletme Araştırmaları Dergisi*, *9*(1), 381–402. doi: 10.20491/isarder.2017.250

Amaro, S., & Duarte, P. (2015). An integrative model of consumers' intentions to purchase travel online. *Tourism Management*, *46*, 64–79. doi: 10.1016/j.tourman.2014.06.006

Bala, M., & Verma, D. (2018). A critical review of digital marketing. *International Journal of Management, IT & Engineering*, *8*(10), 321–339.

Bennett, D., Yábar, D. P.-B., & Saura, J. R. (2017). University incubators may be socially valuable, but how effective are they? A case study on business incubators at universities. In M. Peris-Ortiz, J. Gómez, J. Merigó-Lindahl, & C. Rueda-Armengot (Eds.), *Entrepreneurial universities: Innovation, technology, and knowledge management* (pp. 165–177). Springer. doi: 10.1007/978-3-319-47949-1_11

Buhalis, D. (2003). *eTourism: Information technology for strategic tourism management*. Pearson education.

Buhalis, D., & Deimezi, O. (2004). E-tourism developments in Greece: Information communication technologies adoption for the strategic management of the Greek tourism industry. *Tourism and Hospitality Research*, *5*(2), 103–130. doi: 10.1057/palgrave.thr.6040011

Carvalhosa, P., Portela, F., Santos, M. F., Abelha, A., & Machado, J. (2017, April). Pervasiveness in digital marketing – A global overview. In *World conference on information systems and technologies* (pp. 391–398). Springer. doi: 10.1007/978-3-319-56541-5_40

Chaffey, D., & Smith, P. R. (2013). *Emarketing excellence: Planning and optimizing your digital marketing*. Routledge.

Chen, C. F., & Tsai, M. H. (2008). Perceived value, satisfaction, and loyalty of TV travel product shopping: Involvement as a moderator. *Tourism Management*, *29*(6), 1166–1171. doi: 10.1016/j.tourman.2008.02.019

Demirdögmez, M. (2021). Dijital Pazarlama ve Turizme Yansımaları. In M. Demirdögmez (Ed.), *Modern Pazarlama Yöntemleri ve Turizme Yansımaları* (pp. 1–56). Gazi Yayınevi.

Desai, V. (2019). Digital marketing: A review. *International Journal of Trend in Scientific Research and Development*, *5*(5), 196–200.

Dickinson, J. E., Ghali, K., Cherrett, T., Speed, C., Davies, N., & Norgate, S. (2014). Tourism and the smartphone app: Capabilities, emerging practice and scope in the travel domain. *Current Issues in Tourism*, *17*(1), 84–101. doi: 10.1080/13683500.2012.718323

Gratzer, M., Werthner, H., & Winiwarter, W. (2004). Electronic business in tourism. *International Journal of Electronic Business*, *2*(5), 450–459. doi: 10.1504/IJEB.2004.005878

Happ, É., & Ivancsó-Horváth, Z. (2018). Digital tourism is the challenge of future – A new approach to tourism: Knowledge horizons. *Economics*, *10*(2), 9–16.

Kannan, P. K., & Li, H. (2017). Alice: Digital marketing: A framework, review, and research agenda. *International Journal of Research in Marketing*, *34*, 22–45. doi: 10.1016/j.ijresmar.2016.11.006

Kaur, A., & Sandhu, G. (2017). Opportunities and challenges to digital marketing in developing countries. *International Journal of Science Technology and Management*, *6*(5), 322–327.

Kim, D. Y., Park, J., & Morrison, A. M. (2008). A model of traveler acceptance of mobile technology. *International Journal of Tourism Research*, *10*(5), 393–407. doi: 10.1002/jtr.669

Laajini, T. (2021). Moroccans' behavioural response to the development of digital tourism. *African Journal of Hospitality, Tourism and Leisure*, *10*(6), 1985–1996. doi: 10.46222/ajhtl.19770720.205

Laboy, F., & Torchio, P. (2007). *Web 2.0 for the travel marketer and consumer*. The International Association of Online Communicators.

Law, R., Buhalis, D., & Cobanoglu, C. (2014). Progress on information and communication technologies in hospitality and tourism. *International Journal of Contemporary Hospitality Management*. doi: 10.1108/IJCHM-08-2013-0367

Law, R., Chan, I. C. C., & Wang, L. (2018). A comprehensive review of mobile technology use in hospitality and tourism. *Journal of Hospitality Marketing & Management*, *27*(6), 626–648. doi: 10.1080/19368623.2018.1423251

Mangold, W. G., & Faulds, D. J. (2009). Social media: The new hybrid element of the promotion mix. *Business Horizons*, *52*(4), 357–365. doi: 10.1016/j.bushor.2009.03.002

Mariani, M. M., Buhalis, D., Longhi, C., & Vitouladiti, O. (2014). Managing change in tourism destinations: Key issues and current trends. *Journal of Destination Marketing & Management*, *2*(4), 269–272. doi: 10.1016/j.jdmm.2013.11.003

Murphy, H. C., Chen, M. M., & Cossutta, M. (2016). An investigation of multiple devices and information sources used in the hotel booking process. *Tourism Management*, *52*, 44–51. doi: 10.1016/j.tourman.2015.06.004

Palos-Sanchez, P., Saura, J. R., & Martin-Velicia, F. (2019). A study of the effects of programmatic advertising on users' concerns about privacy overtime. *Journal of Business Research*, *96*, 61–72. doi: 10.1016/j.jbusres.2018.10.059

Pulizzi, J. (2007). *Why content marketing*. https://contentmarketinginstitute.com/2007/04/why_content_marketing/ (Accessed 29 November 2022).

Reino, S., Alzua-Sorzabal, A., & Baggio, R. (2013). What matters to the industry? An evaluation framework for the adoptability of online tourism distribution platforms. In *Information and Communication Technologies in Tourism 2013: Proceedings of the International Conference in Innsbruck, Austria, January* 22–25, 2013 (pp. 412–423). Springer Berlin Heidelberg.

Sawicki, A. (2016). Digital marketing. *World Scientific News*, (48), 82–88.

Sharma, A., Sharma, S., & Chaudhary, M. (2020). Are small travel agencies ready for digital marketing? Views of travel agency managers. *Tourism Management*, *79*, 104078.

Sismeiro, C., & Bucklin, R. E. (2004). Modeling purchase behavior at an e-commerce web site: A task-completion approach. *Journal of Marketing Research*, *41*(3), 306–323. doi: 10.1509/jmkr.41.3.306.35985

Wang, D., & Fesenmaier, D. R. (2013). Transforming the travel experience: The use of smartphones for travel. In *Information and communication technologies in tourism 2013* (pp. 58–69). Springer. doi: 10.1007/978-3-642-36309-2_6

Wang, D., Park, S., & Fesenmaier, D. R. (2012). The role of smartphones in mediating the touristic experience. *Journal of Travel Research*, *51*(4), 371–387. doi: 10.1177/0047287511426341

Wierenga, B. (2010). Marketing and artificial intelligence: Great opportunities, reluctant partners. In *Marketing intelligent systems using soft computing* (pp. 1–8). Springer. doi: 10.1007/978-3-642-15606-9_1

Wi-Fi Alliance. (2022). *2021 annual report*. www.wifi.org/download.php?file=/sites/default/files/private/WiFi_Alliance_2021_Annual_Report.pdf (Accessed 1 January 2023).

Yasmin, A., Tasneem, S., & Fatema, K. (2015). Effectiveness of digital marketing in the challenging age: An empirical study. *International Journal of Management Science and Business Administration*, *1*(5), 69–80.

6

"AIRBNB IRELAND" AS AN "INFORMATION SOCIETY SERVICES" PROVIDER AND NOT A REAL ESTATE OR TOURIST SERVICES PROVIDER

Anastasios E. Thomaidis

Introduction

In the recent decades, the world economy has been undergoing profound changes. There is no doubt that the development of the internet was an important factor in the change in the consumer's behavior.

In particular, the combination of people's concerns and easier access to the internet have created the right moment to build online platforms that are needed to link those who own goods they are willing to share and consumers wishing to use goods. Sharing via platforms [1] is a new model for selling goods and services, which is rapidly developing in various economic sectors, including the tourism sector.[2]

It is true that the travel landscape has already evolved substantially since the dawn of the internet in 1991. In the travel space, Expedia and Booking.com launched in 1996 and the Chinese Alibaba and Ctrip launched in 1999. Online travel agents brought new opportunities for hotels, but also disrupted the traditional relationships between hotels, travel agents, and global distribution systems (GDS). Younger companies like Airbnb, Vrbo, Fpipkey, Agoda, HomeAway, etc., which are based on a peer-to-peer business model, are now stepping forward as the next wave of disruption in the lodging industry.

The collaborative economy, being flexible and focused on the individual, created great opportunities for consumers, businesses and citizens. The tourism sector can easily fit to the collaborative economy, especially in the accommodation and the transport area. Concerning accommodation, platforms offer people a place to stay more affordable than the hotel's prices and the conditions are even better in comparison to an average hotel. In addition, the common dwelling is also rated because it allows access to single accommodation units or has a more authentic experience.[3]

DOI: 10.4324/9781032633374-7

Platforms are digital bilateral or multilateral markets where two or more user groups communicate online with the mediation of the platform operator in order to enable the two groups to enter into a transaction.

In the case of short-term rentals in particular, digital platforms mean any platforms which provide specialized tools for entering into rental agreements online and are not limited merely to promoting the property. On those platforms, natural or legal persons place unused property or unused parts thereof for rental at a pre-agreed price to interested parties (usually tourists) who are seeking to stay at a specific destination. If a transaction is entered into, payment is made online and the company which owns the platform, withholds a small percentage of the transaction amount as a fee/commission, and pays the rest to the owner.[4]

There are several factors that can explain the rise in popularity of the sharing economy in the tourism sector. Digital platforms have made it possible to extend these small exchanges to relations between strangers at an unprecedented scale by reducing the cost of access to the market for individual providers, as well as transaction costs. Specifically, peer-to-peer platforms allow consumers to easily compare prices between different suppliers, find out more about the product or service being offered, review other users' opinions, and, in many cases, communicate directly with the provider of the product or service.

Furthermore, cultural changes and economic developments that have taken place in recent years have led consumers to be increasingly open to the idea of sharing resources and accessing goods on a temporary basis, rather than owning them. In addition, tourist consumers – especially young people – are more open to organizing their trips themselves and increasingly demand unique and personalized experiences. This growing desire to enjoy "authentic" experiences has led to the emergence of new niche markets. In this context, the sharing economy has gained popularity by offering the possibility to engage in these types of experiences, which are more flexible and less standardized, such as stays in unusual places and shared dining experiences with local residents. The model allows consumers to conclude transactions with suppliers in order to secure the products and services they need, which are made available via specially designed online platforms.[5] It is a new form of entrepreneurship worth tens of billions of dollars globally.

The explosive development of online booking of short stays in apartments or houses has created opportunities and opposition in many European capitals. Positive and negative aspects of short-terms rental via digital platforms are a worldwide reality.

Undoubtedly, peer-to-peer business model on which Airbnb and other companies are based on is very simple and profitable for the user and the property owners. The users have the opportunity to organize and personalize their stay in authentic, local properties providing them with a simple – and often less expensive – alternative. On the other side, short-term rentals can be for the owners a substantial source of passive income that requires relatively little effort and the initial investment is often negligible. Staying tourists in real neighborhoods throughout each city can help to improve ugly or deprived areas of the city centers. Short-term rentals normalize prices of accommodations during big events. They also distribute tourism spending. While hotels are located in central business districts, those touristic rentals are scattered usually throughout each city. Finally, the revenue generated by short-term rentals is reinvested in local communities. Whereas hotel profits may go directly back to headquarters, hosts keep money in the area. In the case of Airbnb, hosts retain approximately 87% of the booking price.

On the other side, the rise of short-term rental bookings has come with its fair share of complaints and incidents, among them overtourism, a significant burden on the infrastructure of the cities, skyrocketing rents, and out-of-control parties. Moreover, short-term rentals are increasingly singled out as an aggravating factor in the housing crisis occurring across many countries.

The Airbnb platform

Airbnb[6] is recognized as one of the most successful online sharing accommodation platforms.[7] It has introduced to the tourism and hospitality sectors one of the most significant innovations of the past decade. By facilitating an online peer-to-peer trading platform, Airbnb has extended short-term accommodation capacity in new ways and provoked strong responses from both the established commercial tourism accommodation sector and local regulators.

Airbnb was started in 2008 by Brian Chesky and Joe Gebbia, two industrial designers who recently moved to San Francisco. Unable to afford the rent for their loft at the time, the pair decided to make up the money they needed by renting out their apartment to people who could not find hotels to stay at while attending nearby trade shows. They set up air mattresses in the apartment's living room for their guests to sleep on and cooked homemade breakfast in the morning. Since then, Airbnb has become one of the trailblazers of peer-to-peer property rental.

As of December 2021, the company has thousands of employees. The company operates in 220 countries and almost 100,000 cities and provides its services to the users with its easy user experience platform. The hosts of the company are categorized into two parts, individuals and professionals. They find Airbnb as a stress-free way of earning income from their property. Guests prefer Airbnb over hotels and resorts as they have a cozier and more home-like feeling and are relatively cheaper.

The Airbnb platform is easy and simple both for the host and the guests. The hosts have four ways to list their space on Airbnb. They can list it as an entire home, a hotel room, a shared room, or a private room with an in-depth description, along with multiple photos, pricing and location details, and a description of themselves. After each stay, reviews and ratings from previous guests of both the host and the home are displayed on each listing page. Ratings and reviews on Airbnb are double-blind, which means that both the host and guest cannot see what the other party has written until they have both submitted their feedback. To make a booking, travelers generally need to take the following steps: 1) open the Airbnb website, 2) log in or create an account, 3) specify the location and date(s), using the site's various filters to customize search, 4) make a booking or reserve it – sometimes bookings will not be fully validated until the host accepts, and 5) pay for the accommodation and receive notification of the booking, including the address.

Regardless of what type of property is booked, travelers have the opportunity to message the host before and during the guest's stay. This online conversation creates a strong sense of community and a highly effective support network for travelers. Hosts also generally like to provide local knowledge to help their guests get comfortable in their neighborhoods and homes. Airbnb charges guests a service fee between 5% and 15% of the booking and hosts 3%, generating in 2021, $6 billion in service fees by charging an average of 13% on an average booking value of $156.

In many State Members of the EU, short-term rentals created a confusion concerning legislation. In Berlin, Barcelona, Brussels, Paris, Amsterdam, and elsewhere, Airbnb and other platforms have generated stiff local resistance to the conversion of homes that used to be for renting into de facto tourist accommodation. In turn, this has led authorities to take measures to counter a development that has shown its potential to change the face of cherished parts of cities, creating several legal crises.

In response the companies behind the platforms such as Airbnb, Home Away, and others – plus the trade associations they belong to, in this case, the European Holiday Home Association – have launched a lobbying offensive in the EU institutions. This includes a complaint against four cities in particular, Barcelona, Berlin, Paris, and Amsterdam, that the local government's attempts to regulate the tourist rental market breach the EU's single market rules, cases which ultimately ended up at the European Court of Justice. The online rental platforms are enlisting the European Commission's help to roll back defensive measures against them taken at the local level.

The case law of the CJEU relating to short-term rentals (C-390/18)

Although Airbnb is popular amongst travelers particularly in Paris and all over France, the French government installed, under pressure from the hotel lobby, a range of measures to limit the market growth of Airbnb, including a tourist tax and a limit of the duration of rentals. Besides these measures, the Association pour un hébergement et un tourisme professionnels ("AHTOP") lodged a complaint concerning the classification of Airbnb as a real estate agent.[8] If classified as such, Airbnb would need a license for its activities. The dispute following the complaint resulted in a preliminary reference to the Court of Justice of the European Union ("CJEU") on the applicability of Directive 2000/31 and the compatibility of the relevant French measures with this Directive. Put simply, the CJEU had to decide whether Airbnb provides an information society service or an accommodation service. Consequently, the Airbnb case provided the CJEU with the opportunity to clarify or further cement the conclusions drawn in previous cases with regard to the concept of "information society services."

More specifically by its first question, the Regional Court of Paris asked whether the services provided by Airbnb are information society services within the scope of Directive 2000/31. Secondly, if the CJEU decides that Airbnb falls within the scope of the Directive, the question arises whether the Hoguet Law can be enforced against Airbnb.

The AHTOP argued that Airbnb's services are accommodation services and fall under the definition of services in the Services Directive that sets forth the free movement of services in the broad sense. Additionally, they argued that Airbnb's services do not qualify as information society services in the sense of Directive 2000/31. The AHTOP argued in subordinate order that the platform services form an integral part of an overall accommodation service.[9]

By contrast, Airbnb denied being a real estate agent and hinged its argumentation on the alleged incompatibility of the Hoguet Law with Directive 2000/31. This Directive guarantees the free movement of information society services by forbidding Member States to introduce barriers to the free movement of information society services provided by a company that is established in another Member State. In that regard, an information society service is defined as "any service normally provided for remuneration, at a distance, by electronic means and at the individual request of a recipient of services." Directive 2000/31, however, allows Member States to derogate from the general rule and to adopt measures restricting the freedom to provide information society services if they are necessary for extraordinary objectives, such as public health or public security, and they are taken against a company violating or risking to violate those objectives. However, the Member State in question is obliged to first ask the Member State of the place of establishment of the offending company to take measures and it has to notify the European Commission. Only in case of urgency, a Member State is allowed to take measures without notifying the European Commission and the Member State of the place of establishment. None of the parties has argued that there was an urgency in this given case scenario. First, the CJEU assessed whether the services provided by Airbnb are information society services. The CJEU described Airbnb's services as follows: "an intermediation service which, by means of an electronic platform, is intended to connect, for remuneration, potential guests with professional or nonprofessional hosts offering short-term accommodation services, while also providing a certain number of other services." The CJEU emphasized that the services provided by Airbnb, either way, fall within the definition provided by Article 57 TFEU and the Services Directive. However, Article 3 of the Services Directive determines that other, more specific directives take precedence in case of a conflict with the provisions of the Services Directive. This called for an analysis of whether Airbnb's services fall within the scope of Directive 2000/31 and are thus information society services. Several cumulative conditions need to be fulfilled in this regard. First, the CJEU decided that the Airbnb services are provided for remuneration, even though this remuneration is only paid by the guest. Second, the services are supplied electronically and at a distance, because Airbnb and the host or guest exchange information only on the platform. Finally, the service is granted at the individual request of the recipients. The CJEU thus, preliminary, concluded that the Airbnb services are information society services, and fall within the scope of Directive 2000/31. The analysis of the CJEU did not stop here. It recalled the conclusion reached in the Asociacion Profesional Elite Taxi case ("Uber Spain"), where the CJEU concluded that if the service "forms an integral part of an overall service whose main component is a service coming under another legal classification", this service cannot be classified as an information society service differing from the overall service to which it is linked. Somewhat confusingly, the CJEU had previously decided that the service of Airbnb was provided electronically and at a distance, while the argumentation about mixed services is triggered by the fact that the service is not provided fully electronically or at a distance, as was the case in the Ker-Optika and Uber cases. If the CJEU wanted to be fully consistent with the line of argumentation in previous cases, it could have stopped its analysis after this conclusion.

Nonetheless, the CJEU analyzed whether the platform services form an integral part of an overall service, more specifically an accommodation service. The CJEU decided that the essential element of the service of Airbnb is the listing of the accommodations and, thus, that the intermediary service cannot be seen as ancillary to another service. In his opinion, the Advocate-General deduced two conditions from the Uber cases for the conclusion of an inseparable link between the accommodation service and the intermediary service: (i) the services create a new market and (ii) the provider exerts decisive influence. If the CJEU finds an inseparable link, the intermediary services would be an integral part of the accommodation service and thus be judged as such. The CJEU followed this line of argumentation and emphasized, first, that Airbnb did not create a market for accommodation services. In that regard, the CJEU mentioned estate agents, classified advertisements and property letting websites, who created this market long before Airbnb was founded. With regard to the creation of a market, the Advocate-General rightly pointed out that Airbnb, contrary to Uber, also allows professional hosts, for whom there was already a market, on its platform. The CJEU did, however, not include this argument in its judgment. Second, the CJEU highlighted the free choice of hosts in determining the renting price, which shows a lack of decisive influence on Airbnb's part. Taking all this into account, the CJEU concluded, in line with the opinion of the Advocate-General, that the platform services are not an integral part of an overall accommodation service and are thus information society services. The CJEU then further analyzed whether the ancillary services, such as the photography service, changed the classification of the services. In this regard, the CJEU emphasized again that the host sets the price and points out the paradoxical nature of ancillary services changing the classification of the intermediation service. Thus, it concluded that the ancillary services do not substantially modify the nature of the intermediation service.

The CJEU continued by explaining why a different conclusion was drawn for Airbnb than in the two Uber cases. The crux of the argumentation in the Uber cases was whether Uber exercised decisive influence, which was held to be the case. Contrary to Uber, Airbnb does not exercise such decisive influence in the CJEU's opinion, because it does not decide the rental price and it does not, to a large extent, select the host or the accommodation allowed on the platform. Because the CJEU answered the first question affirmatively, it proceeded to analyze whether the Hoguet Law is enforceable against Airbnb and thus compatible with Directive 2000/31. The CJEU first confirmed the restrictiveness of the Hoguet Law. The measure violates the free movement of services because it restricts access to an activity through the requirement to hold a license. However, Member States are allowed to derogate from the free movement of services if they fulfill the requirements of Article 3(4) of Directive 2000/31. Contrary to the Advocate-General, the CJEU did not analyze the first prong of the derogation, that is whether the measures are necessary and proportionate. The second prong of the derogation requires the deviating Member State to request the Member State where the undertaking is established to undertake steps. In this case, the French government, as the deviating Member State, refrained from doing so. The question then remains whether this has an impact on the enforceability of the Hoguet Law. The CJEU first pointed out that the Directive is "sufficiently clear, precise and unconditional to confer on it direct effect" Secondly, the CJEU confirmed that the requirement to contact the Commission and the other Member States is an essential procedural requirement. Finally, the CJEU emphasized that the requirements in the Directive are there to stop Member States from acting beyond the competencies laid down by European Union law. In conclusion, the CJEU deemed the Hoguet Law unenforceable against individuals based on its incompatibility with Directive 2000/31.

To sum up, in its judgment of December 19, 2019, in the Airbnb Ireland case,[10] the Court ruled that an intermediation service (which, by means of an electronic platform, is intended to connect, for remuneration, potential guests with professional or non-professional hosts offering short-term accommodation services, while also providing a certain number of services ancillary to that intermediation service) must be characterised as an "information society service" falling within the scope of Directive 2000/31/EC on electronic commerce.[11]

In addition, the European Court of Justice ruled in two cases (C-724/18; C-727/18) concerning short-term rentals that national or local requirements to register short-term letting are not in violation of European law.

In the key thesis of the judgment in the case of Cali Apartments SCI and HX against Procureur général près la cour d'appel de Paris i Ville de Paris (C-724/18 i C-727/18), the Grand Chamber of the Court of Justice ruled that national legislation providing for authorization for short-term rental to tourists is consistent with the EU law. It has deemed that combating the shortage of long-term housing represents an overriding reason of public interest, and therefore, it provides grounds for implementation of legal restrictions on short-term rental.

The Court of Justice examined the issue of authorization that requires from the short-term rental operator to take certain steps and from the relevant authorities to issue a formal decision granting him permission for operations.[12] Another issue that was raised concerned the overriding reasons relating to the public interest[13] on which Member States may rely. According to the ECJ case law, they are: the protection of the urban environment,[14] the aims of social policy, and, last but not least, the shortage of housing for long-term rental. ECJ held in the judgment that the service activity consisting in repeated, short-term renting out furnished residential units to tourists does not fall within the definition of "services" outlined in Article 4 (1) of -Directive 2006/123/EC and it is not listed among the activities that are excluded from the application of the Directive under the provisions of the Article 3 (2). The judgment confirms that the provisions imposing the obligation to obtain an administrative decision of a competent authority granting a service provider a permission to operate fall within the scope of the definition of an authorization scheme outlined in Article 4, Paragraph 6 of the said Directive. Paragraph 109 of the Judgment indicates that the Directive 2006/123/EC applies to the national regulations of a Member State that pertain to business operations consisting of repetitive, short-term rental of furnished accommodation to transient clientele who do not take up residence there. Furthermore, the Services Directive grants national authorities the power to establish, within the framework set out in that regulation, the conditions of authorization in light of the objectives of social diversity, taking into account the characteristics of local housing markets and the need to tackle the problem of the housing shortage.[15]

Conclusions

The emergence of online intermediary platforms has fostered an unprecedented rise in short-term rentals across Europe, rendering them a considerable alternative option for tourist accommodation. In particular, Airbnb and other similar digital platforms have brought numerous benefits to property owners and tenants who have unoccupied rental units. The Airbnb platform has also created a boom in local and international tourism, increasing taxes for the government. It has employed thousands of workers in different parts of the world. Additionally, the agency acts as a trustworthy tax collector for the government. On the other hand, the company has created problems, such as cases of illegal listing and rising costs of housing. Specifically, this development has brought about tensions in European cities related to housing shortage, disturbance to neighborhoods, and competition with traditional accommodation providers. Furthermore, the situation poses significant issues in various areas of European law, including consumer protection, internal market, data protection, and taxation.

However, the benefits of home-sharing and short-term rentals cannot be overlooked. Cities and local governments should understand that completely banning short-term rentals is not the way to go. Through technology, several cities have been able to oversee and regulate short-term renting. American cities like Colorado and North Carolina have successfully utilized technology to enjoy compliance and growth in revenue collection. These cities have also outlined workable regulations for hosts and tourists that have been quite successful, and it could be an example for European countries.

In light of the case law of the CJEU, platforms such as Airbnb cannot be understood in many legislation systems as providing "real estate agency"-services nor as falling within the scope of the specific legal framework that lays down the terms and conditions for carrying on the profession of real estate agent. It is therefore not possible to impose on Airbnb the relevant criminal and disciplinary sanctions that are threatened.

In all events, this case law entails the risk that any restrictions which are considered to be unacceptable under European law are found incompatible with European law, with the result that they would not be implemented by the national judge. The absolute nature of the case law position supports the general prohibition on the adoption of restrictions, without any concern as to whether the criterion to be used relates to the lessor, the duration of the rental, the geographical region in which the property is located, the income the lessor derives from exploiting his/her assets, or the justification for introducing restrictions.

Furthermore, the ECJ judgment imposes an obligation on national courts to carry out a comprehensive assessment of the situation in this area in order to meet new challenges. It provides an opportunity for national legislators to regulate short-term rental in a way that ensures fair competition and protection of customers. The postulated solution is to adopt appropriate legal regulations at the level of national law, which would, above all, ensure effective protection of consumers, and to consider adopting appropriate solutions at the local level, which would enable local authorities to influence the development of short-term rentals.[16]

Especially, in Europe legal certainty could also be achieved if the legislator creates a legislative framework regarding the services provided by platforms. It has generally been accepted that

"two of the most significant obstacles to the development of the digital economy have been the existence of outdated national legislation and regulation which was enacted decades ago and the regulatory uncertainty regarding the rules, rights and obligations applicable to digital platforms."

The most efficient way to regulate online intermediaries would be to adopt legislation at the European level.[17]

Moreover, this decision is important also for the legal nature of other similar platforms such as Bookin.com.

Notes

1 Tassikas, A. & Thomaidis, A. *Regulation of short-term tourist rental in Greece in Apollonia Martinez Nadal*, Short-Term Tourist Rental Regulation a Comparative Law Approach, p. 88.

2 In Greece, this is related particularly to the accommodation sector and other services (tours and excursions). See Thornton, G. (2019, February). *Survey for the hellenic chamber of hotels*, p. 2
3 Stiubea E. (2018) University of Oradea , *Romania, new tendencies in tourism: The sharing economy*, p. 239.
4 Tassikas, A. & Thomaidis, A. *Regulation of short-term tourist rental in Greece in Apol-lonia Martinez Nadal*, Short-Term Tourist Rental Regulation a Comparative Law Approach, p. 88.
5 Busch, C, (2019), Mehr Fairness und Transparenz in der Plattformökonomie? Die neue P2B-Verordnung im Überblick,. *Gewerblicher Rechtsschutz und Urheberrecht* (GRUR), 8,788–796 (passim).
6 Air Bed and Breakfast.
7 Sthapit, E., Coudounaris, D. N., & Björk, P. (2019a). Extending the memorable tourism experience construct: An investigation of memories of local food experiences. *Scandinavian Journal of Hospitality and Tourism*, *19*(4/5), 333–353.
8 France, one of the EU Member States where Airbnb is active, passed the Hoguet Law in 1970. Under this act, all real estate agents are required to hold a professional license under penalty of 6 months imprisonment and a fine of 7500 euros.
9 Van Acke, L. C-390/18 – The CJEU *f*inally *cl*ears the *A*ir(bnb) *r*egarding *i*nformation *s*ociety *s*ervices – beck-online, p. 1, (file:///C:/Users/%CE%96%CE%A9%CE%97/Downloads/C-390_18%20%E2%80%93%20The%20CJEU%20Finally%20Clears%20the%20Air(bnb)%20Regarding%20Information%20Society%20Services%20-%20beck-online%20(1).pdf).
10 ECJ decision from 19.12.2019, C-390/18 Airbnb Ireland UC, Hôtelière Turenne SAS, Association pour un hébergement et un tourisme professionnels (AHTOP), Valhotel, ECLI:EU:C:2019:1112.
11 Since the conditions in Article 1(1)(b) of the Directive (EU) 2015/1535, which refers to Article 2(a) of Directive 2000/31/EC, are met. See recitals 49, 69 of the judgment.
12 See Judgment of the CJEU, Case C-360/15 and Case C-31/16, X and Visser, EU:C:2018:44, and also Judgment of the CJEU, Case C-393/17, Kirschstein, EU:C:2019:563.
13 See Judgment of the CJEU, Case C-567/07, Woningstichting Sint Servatius, EU:C:2009:593, and also Judgment of the CJEU, Case C-197/11 and C-203/11, EU:C:2013:288.
14 See point 135 Judgment of the CJEU, Case C-360/15 i C-31/16, X and Visser, EU:C:2018:44.
15 Badura, E *The judgment of the European Court of Justice of 2022 C-724/18 and its impact on the real estate market in the EU*, pp. 87–89, https://hrcak.srce.hr/file/375778 (accessed 1 September 2022).
16 Badura, E. *The judgment of the European Court of Justice of 2022 C-724/18 and its impact on the real estate market in the EU*, pp. 87–89, https://hrcak.srce.hr/file/375778 (accessed 1 September 2022).
17 Van Acke, L. *C-390/18 – The CJEU finally clears the Air(bnb) regarding information society services – beck-online,* p. 2, (file:///C:/Users/%CE%96%CE%A9%CE%97/Downloads/C-390_18%20%E2%80%93%20The%20CJEU%20Finally%20Clears%20the%20Air(bnb)%20Regarding%20Information%20Society%20Services%20-%20beck-online%20(1).pdf).

References

Badura E. (2021) The judgment of the European court of justice of 2022 C-724/18 and its impact on the real estate market in the EU. *Zbornik Pravnog fakulteta u Zagrebu*, *71*(1).

Busch, C, (2019), Mehr Fairness und Transparenz in der Plattformökonomie? Die neue P2B-Verordnung im Überblick. *Gewerblicher Rechtsschutz und Urheberrecht* (GRUR), *8*, 788–796 (passim).

Sthapit, E., Coudounaris, D. N., & Björk, P. (2019a). Extending the memorable tourism experience construct: An investigation of memories of local food experiences. *Scandinavian Journal of Hospitality and Tourism*, *19*(4/5), 333–353.

Stiubea E. (2018). *New tendencies in tourism: The sharing economy*, p. 239.

Tassikas, A., & Thomaidis, A. (2022). *Regulation of short-term tourist rental in Greece in Apol-lonia Martinez Nadal*. Short-Term Tourist Rental Regulation a Comparative Law Aapproach.

Thornton, G. (2019, February). *Survey for the hellenic chamber of hotels*, p. 2. www.grant-thornton.gr/press-releases/2019/sharing-economy-2019/ (Accessed 30 November 2021).

Van Acke, L. C-390/18-*The CJEU finally clears the air(bnb) regarding information society service-Beck-online*, p. 1.

PART 2

Knowledge sharing

Case studies

7

DESTINATION MANAGEMENT. THE ROLE OF DMOs ON DIGITAL BRAND COMMUNICATION. THE CASE OF GREECE

Panagiota Dionysopoulou and Eftychia Christina Aivaliotou

Literature review

There are only a few references, surveys and papers on tourism governance and the digital presence of DMOs worldwide. This limitation on sources and resources makes difficult the comparative analysis among countries. Nevertheless, the best outcome is produced by taking into account the data limitations. In order to better comprehend the contribution of the research on this topic, we have gathered the most essential ones in Table 7.1.

Most of the above-mentioned surveys and research papers refer mainly to the DMOs' contribution to social media advertising campaigns, to the destination Logo recognition or to the determinants of their success. Little is known about their key purpose of establishment, namely their contribution to policy planning, governance and effective destination management. Since there is such research limitation, future studies should focus on a further analysis of the necessary factors, provide more case studies or best practices and call on legislation reforms that will enable the establishment and implementation of such structures at the national and local governmental levels.

Research context and methodological framework

The methodology of our research is based on a comparative analysis, depending on the available data. Greece is examined as a case study, as the country – during the last years – has made essential steps toward this direction. The key question of this paper is to showcase the best way DMOs can be involved in the governance of a destination. "How can policy makers set up and implement a more efficient digital brand communication at national and local level taking into account all the

DOI: 10.4324/9781032633374-9

TABLE 7.1 Contribution of previous studies

Reference	*Subject*	*Main content/findings*
Pietro Beritelli and Christian Laesser (2018)	Destination Logo recognition and implications on international destination branding by DMOs: a case for saving money	This research paper indicates that destination logo recognition is very limited and it is primarily useful for the product in the destination, rather than for destination communication. Therefore, the impact of branded communication and advertising campaigns is greatly overestimated.
Lidija Lalicic et al. (2018)	Emotional Brand communication on Facebook and Twitter: Are DMOs successful?	This paper analyzes the emotional brand communication of the 10 most popular DMOs in Europe according to Tripadvisor (2017) by observing user responses across Facebook and Twitter. It provides DMOs with guidelines on how to effectively communicate their brands using specific emotional brand values on social media and suggests values that DMOs should integrate into their social media communication strategies regarding their destination brands.
Sebastian Molinillo et al. (2018)	DMO online platforms: Image and intention to visit	It explores the moderating effects of DMO online platforms on the destination image (DI) through a conceptual model. The findings demonstrate that tourist involvement has a positive impact on the cognitive image and affective image, forming the DI as an antecedent of the intention to visit. The results also indicate that image formation and intention to visit the destination vary depending on the platform used by travelers to access the information.
Tom Bornhorst et al. (2010)	Determinants of tourism success for DMOs & destinations: An empirical examination of stakeholders' perspectives	This paper examines the concept of tourism success as it relates to both destinations and to DMOs, and determines whether a relationship or distinction exists between these two. Unique to DMO's success were supplier relations, effective management, strategic planning, organizational focus and drive, proper funding and quality personnel. Unique to destination success were location and accessibility, attractive product and service offerings, quality visitor experiences and community support.
Ahmet Uşaklı (2017)	How "social" are destinations? Examining European DMO social media usage	The study explores how European DMOs use social media to promote and market their destinations and extends extant research on social media in tourism. Findings suggest that European DMOs utilize social media as a traditional marketing tool, rather than exploiting it as a customer service tool to mitigate potential customer problems.
Michael Volgger and Harald Pechlaner (2014)	Requirements for destination management organizations in destination governance: Understanding DMO success	This paper aims to better understand DMO and destination success by investigating the role played by networking capability. One key insight is that the effect of networking capability works through increased DMO authority, that is enhanced power and acceptance of the DMO within the destination network.

relevant stakeholders and structures such as DMOs?" More precisely through the comparative analysis, this paper showcases the best case scenarios and highlights the most efficient practices of tourism policies and governance that have actually achieved their main objectives and mission. The key task of this research is the conclusion on the necessary features, elements and components a tourism policy should comprise in order to be successful and effective. Emphasis is given on the role of DMOs and their contribution at either a regional or a national level of governance and to the digital brand communication of a destination. The comparative analysis of different regions and policies will assist policy makers by indicating which policy making process is the most appropriate for the chosen model of tourism governance.

Comparative analysis

First of all, it is important to set the definition and the mission of DMOs according to the guidelines of the Institutional Strengthening of Destination Management Organizations of UNWTO (UNWTO, 2019). During the last years, we have been witnessing a shift from traditional marketing national tourism boards to DMOs as leading organizations assigned with a variety of tasks, such as strategic coordination and management of a range of activities and different stakeholders sharing a common vision: the effective destination management. According to UNWTO, destination management is "the coordinated management of all the elements that make up a tourism destination (attractions, amenities, access, marketing and pricing)" (UNWTO, 2019). Destination management takes a strategic approach to link up the sometimes very isolated and occasionally diverging elements for the better planning and management of the destination. It is very essential that DMOs align their actions and plans with the UNWTO Global Code of Ethics for tourism, the SDGs of UN Agenda 2030 and especially with the SDGs 8: decent work and economic growth, 12: sustainable consumption and production and 14: life below water – that are related with the tourism sector. Monitoring mechanisms on their implementation should ensure their compliance with the ultimate goal.

The DMO's role should be to lead and coordinate activities under a coherent strategy in pursuit of this common goal. DMOs do not control the activities of their partners and other actors operating in the tourism destination, but they bring together resources and expertise, and a degree of independence and objectivity to lead the way forward. Though DMOs have typically undertaken marketing activities, their remit is becoming far broader: being a strategic leader in destination planning and management. This is a vital ingredient for their success. DMOs are increasingly planning and implementing strategies aiming at improving the competitiveness of their destinations through an enhanced and responsible management of their natural and cultural attractions which ensures the destination's sustainability in the long term, at the creation of additional innovative resources and solutions, and at enhancements in the efficiency of their production and distribution processes. The ultimate goal is to stimulate sustainable consumption and production and facilitate visitors' interaction with the destination and its residents.

A step further in destination management is the concept of the smart destination, which is founded on five pillars:

1 Governance
2 Innovation
3 Technology
4 Accessibility and
5 Sustainability

The five pillars and the interaction among them are the cornerstones for the transformation of a destination into a smart destination, and the DMO is at the center of this transformation. A DMO is "the leading organizational entity which may encompass the various authorities, stakeholders and professionals and facilitates partnerships towards a collective destination vision."[1] The governance structures of DMOs vary from a single public authority to a public–private partnership model – to a lesser extent also entirely private models are found – with the key role of initiating, coordinating, and managing certain activities. As the leading organizational entity in the destination, the DMO needs to get the support of all acting stakeholders and to be considered as the legitimized entity – institutionally, legally, and operationally speaking – to lead the management of tourism in the destination. They should guarantee a sustainable management of the destination, under an environmental, social, and economic approach. There are many different roles and responsibilities as far as destination management is concerned. In the event that in certain destinations there are not yet DMOs established, some fundamental DMO's functions, when created, may include the following:

- Strategic planning
- Design (or participation in the designing process) and implementation of the destination's tourism policy
- Market intelligence (data gathering and analysis, market research, etc.)
- Tourism product and business development
- Digitalization and innovation
- Monitoring
- Crisis management
- Training and capacity building (not only of its human resources but also facilitate training and capacity building activities for local tourism professionals)
- Promotion, marketing and branding
- Funding and fostering investments

(UNWTO, 2019)

In order for DMOs to be successful, they should aim at inclusive growth that benefits both local businesses and local communities as well. They should be able to raise the awareness of the socioeconomic benefits of the tourism sector and therefore engage all the relevant stakeholders in regards to sustainable tourism development. Of course crucial is their contribution to building a strong brand destination identity. The value system that will be incorporated will result in repeated visitors and their connection with the destination. The assessment and evaluation process on a regular basis is a significant tool that will result in the improvement of its performance key areas. Their strategic leadership refers to the following issues:

- Coordination
- Policy
- Values and
- Vision

The DMOs should strengthen their internal capacities in three key performance areas:

- Strategic leadership
- Effective execution and
- Efficient governance

The above-mentioned indicate that in order for a DMO to be able to lead on strategic planning, it is imperative to have certain aspects such as reliability, capacity to engage, authority, responsibility, legitimacy, vision, decision-making and recognition. The effective implementation requires an operational plan, implementation, measurement, management, coordination, KPIs, monitoring, and strategic planning, while effective governance requires transparency, accountability principles, public–private partnerships, adequate structure, cross-cutting approach, public consultations, inclusive growth, and alignment with SDGs. Their activity should focus on specific areas (Tom Bornhorst et al., 2010):

1. Coordination of different tourism actors for a unified voice and representation
2. A leadership and advocacy role within the local community it services
3. Development of the destination image in order to be competitive
4. Provision of visitor services and further information upon their arrival
5. Liaison with TOs, travel agents and MICE representatives that bring travelers to the destination

The determinant factors of their success relate to community support, marketing and destination performance provided stable political and economic conditions. Furthermore, networking, transparency, resources and professionalism are important actors of a DMO's success, which is strongly linked with the success of a destination (Volgger & Pechlaner, 2014). Once established, the next step for DMOs is to create a social account since the digital presence reaches a greater audience than with the traditional means of communication and promotion (Usakli et al., 2017).

Especially, in **Europe** it is in their best interest. Determinant factors for an effective social media usage are as follows: content, major theme, information type, engagement, interactivity, promotion, and customer service. These categories should be taken into account when measuring the social performance of the DMOs.

Through the comparative analysis we carried out, we came up with some interesting findings on the digital role of DMOs worldwide. In Japan, the DMO concept is relatively new, while in Europe, it is a rather acquainted notion. Finland, Spain, Netherlands, Ireland, France, Cyprus, Luxembourg, and Norway are among the European countries that have established DMOs and are active in social media as well. Table 7.2 presents in a comprehensive way the most interesting case studies.

Based on the above, we come to realize that there are different missions and tasks DMOs are called to achieve, depending mostly on the top priorities of the country's tourism strategy. Besides the main task, namely the promotion of the brand, there are additional challenges that each destination wants to measure up to. For instance, Zürich has set sustainability and the environmental footprint very high in the tourism Agenda, Berlin's structure is based on a private-public partnership in order to raise funding, while Helsinki has involved the voice of the local community and that has been a major factor of its success. Of course, all of them are active on social media (Twitter, LinkedIn, Facebook, Instagram), where their brand campaigns are promoted.

The case of Greece

In Greece, in 2019, an initiative was taken by the Hellenic Ministry of Tourism that focused on the challenges of tourism development in three of the country's top destinations – Rhodes, Corfu and Santorini – with a view to increasing positive impacts on the tourism sector and local communities. The European Bank for Reconstruction and Development (EBRD) implemented a pilot program from May to October 2019 in cooperation with the Hellenic Ministry of Tourism and the Chambers of Commerce of the Cyclades and Dodecanese, through funding from the European Union.[7] The European Union-funded destination management planning initiative aimed to assist the islands' businesses and local communities in developing responses to the challenges of tourism, including diversifying destinations, reducing the dependence on "Sun and Beach" model and combatting the threat of "over-tourism."

Furthermore, the Hellenic Ministry of Tourism issued in 2021 a new law bill (L. 4875/2021 Gaz. A' 250/2021) with a range of provisions that aim to resolve a number of sectoral issues of economic, social, developmental, and environmental importance. The new bill, which among others includes the establishment of DMMOs, was drawn to reduce bureaucracy and support investment projects in Greece. More specifically, the new bill of the Ministry foresees:

- The establishment of **Destination Management & Marketing Organizations (DMMOs)** to secure a public–private cooperation framework for the strategic

TABLE 7.2 Case studies on DMOs

Country	*Institution*	*Initiative*
Romania	DMO in Brasov County	The Association for the Promotion and Development of Tourism in Brasov County (APDT Brasov)[2] was founded in 2006, within a local development project and created in a public–private partnership format. The Association for the Promotion and Development of Tourism (APDT) Brasov has, as its main objective of the activity, the management of the tourist destination and the unitary promotion of the representative tourist objectives of Brasov County at a national and international level (APDT Brasov) (Foris et al., 2020).
Germany	*visitBerlin*[3] (Berlin Tourismus & Kongress GmbH)	VisitBerlin has been globally promoting Berlin as a brand. VisitBerlin showcases Berlin at hundreds of events globally attracts international journalists and bloggers to the city and supports them on their research trips here. Welcome in Berlin (Willkommen in Berlin) is the key message of their communication. In the convention sector, the **Berlin Convention Office** acquires conferences, fairs and congresses for Germany's capital city. *VisitBerlin* also functions as a **travel agency** offering hotel accommodation and tickets, and issuing the ***Berlin Welcome Card***, Berlin's official sightseeing pass. In addition, it operates the **Berlin Tourist Info Centers** and the **Berlin Service Center**. With its vast range of useful information plus exclusive offers, Berlin's official online **tourism platform**, makes it easy for visitors to prepare their trip to Germany's capital city. *VisitBerlin* is funded by the Berlin Senate as a **public-private partnership**. However, a sizable proportion of its budget is generated by its own activities and reinvested in promotional campaigns for Berlin. *VisitBerlin* is ranked as one of Europe's most successful destination management organizations.
Helsinki	**Helsinki** Marketing, which is a marketing company owned by the City of Helsinki.[4]	Helsinki Marketing is responsible for Helsinki's operative city marketing and business partnerships. Helsinki Marketing's scope of operations covers marketing activities related to tourism, congresses, events and business promotion, application procedures for major events and congresses, as well as tourist information services. Its target audiences include local residents and international tourists, decision-makers and experts. Helsinki Marketing works in close cooperation with the departments and units of the City of Helsinki. Furthermore, all the content of the website is compiled by the locals. MyHelsinki is an attempt to have the locals tell their best stories of the city. Tips at tourists are given through personal and genuine recommendations by locals and not by paid ads or by any sponsors. Their brand communication engages the voice of the local community and this perspective distinguishes this DMO in comparison to others.

(Continued)

TABLE 7.2 (Continued)

Country	*Institution*	*Initiative*
Netherlands	**NBTC** (Netherlands Board of Tourism And Conventions)	As the destination management organization of the **Netherlands**,[5] NBTC provides a national vision for tourism based on relevant knowledge and insights. They use this vision to connect, facilitate and support cities, regions, business owners and other parties; they work closely with the tourism industry, regional and local destination management organizations, provinces and Ministries. This allows them to jointly develop initiatives that attract visitors for holidays or business meetings in the Netherlands. It is an independent foundation with a Supervisory Board. The digital communication of Netherlands includes the element of surprise, adventure, history, memories, relaxing; elements that combined together differentiate the country from other European destinations.
Zürich	**Zürich Tourism**[6]	Zürich Tourism is responsible for destination marketing and thus for profiling the brand of the city and region of Zurich as a highly diverse destination. It promotes **quality and sustainability** and, in collaboration with its partners, is systematically implementing this philosophy. Top positions in international rankings underscore this commitment. Zürich Tourism has been awarded the Quality Label, Level III of the Swiss Tourism Federation (STF). In this way, it intensively endorses the notion of providing top quality in the Zurich tourist region. At the same time, Zürich Tourism actively promotes ecological, economic and social sustainability, and has been leading by example in this respect since 1998. Both its staff and its service providers are wholeheartedly committed to this idea. In 2009, Zürich Tourism signed the Sustainability Charter drawn up by the Swiss tourism industry. The organization has been certified according to ISO 9001 and ISO 14001. In addition, since 2010, Zürich Tourism has been offsetting all non-avoidable CO_2 emissions in collaboration with the myclimate foundation. On doing so, it supports an international climate protection project. In order to promote social sustainability, Zürich Tourism also collaborates with the association "BarrierefreieSchweiz." In line with the measures adopted by both the City (Strategy 2035, Smart City) and Canton (Cluster Strategy, innovation park) of Zurich, in the long term Zürich Tourism is pursuing the vision of becoming a so-called "smart destination." As a destination, Zurich regularly dominates the top positions in global sustainability rankings – such as 1st place in the Arcadis Sustainable Cities Index 2016 or 2nd place in the Mercer Quality of Living Ranking 2019. Furthermore, Zurich is one of the most sustainable congress cities in the world, achieving the third highest number of points in the Global Destination Sustainability Index (GDS). In 2017 Zurich supported the International Year of Sustainable Tourism for Development endorsed by the UNWTO as a "Friend of the Year."

promotion of tourism products and ensure the quality of the services provided. The planned DMMOs are considered to be a key tool for changing Greece's tourism model and finding solutions to issues such as over-tourism in certain destinations.
- The establishment of "Standard Integrated Tourism Management Destinations," which will look to detect the country's destinations of national importance and set a framework for their protection by monitoring capacity indicators, setting development strategies and coordinating the creation of the necessary infrastructure to support each area.[8]

More specifically, according to this regulatory framework, the DMMOs strategic planning focuses on four main points. Firstly, DMMOs, through an integrated policy-making process that focuses on the development of clusters of local and regional authorities and the creation of synergies between the public and private sectors, will promote in a coordinated way the goals of sustainability and sustainable tourism development. Secondly, the DMMOs will play a key role in the promotion of specific tourism products at local and regional level that guarantee a minimal environmental footprint. Thirdly, emphasis will be given to the promotion of green and sustainable practices within the tourism business sector that not only contributes to the objective of sustainability but shall attract customers that value environmentally friendly businesses. Last but not least, particular focus will be given to the promotion of local and regional key thematic sectors that create a unique identity for the destination, differentiate it from its competitors, and enhance its competitiveness in the international tourism market. The new bill defines that the DMMOs aim at the scientific, consulting and technical support of the Local Authorities, at the promotion and accurate information on the tourism product for the respective destination, the set-up, support, and implementation of the development of tourism policy of the destination. The DMMOs will offer support to local and regional administrations for the effective implementation of their tourism policy within the framework of the Strategic Plan for sustainable tourism development. Promotion of the destination brand and market research are core competencies of these organizations. Key objectives will be the monitoring of the upgrade of infrastructure and services of public good, the elaboration and implementation of a Strategic Plan for sustainable tourism development – for management and promotion of the destination, the elaboration of studies for the creation of specific standards for the enhancement of brand destination, the planning and the implementation of the Marketing strategy, the creation and management of local quality labels, the development of research work on important tourism issues, such as market analysis, infrastructure projects and the implementation of policies of social cohesion, digital convergence and sustainable development. In order to fulfill their mission, the DMMOs will collaborate with professional associations, public institutions and business clusters to safeguard the quality of the services offered. They will ensure the quality of the provided tourism product through co-advertising, funding of studies, management and promotion of the destination domestically and internationally. It will be possible for DMMOs to cooperate with specialized Tourism Organisations, Tour Operators, travel agents and airlines for the implementation of joint co-advertising programs at a local level. They may also plan and carry out volunteering and social responsibility actions related to sustainable tourism development and the preservation of the residential and natural environment of the destination and provide incentives for the promotion and management of local quality labels.

Specific destinations, which contribute essentially to the country's brand name, will be nominated as "Standard Integrated Tourism Management Destinations." They will be selected upon criteria related to their international visibility, carrying capacity, natural and cultural resources, and sustainability. In this case, the relevant Local Government must, within a relatively short time, establish a DMMO or amend the existing Special Purpose SociétéAnonyme into a DMMO. The DMMO of a "Standard Integrated Tourism Management Destination" will be obliged to prepare and provide an annual Report on the current situation of the destination, its vision, and its development strategy with reference to quantitative and quality indicators, and to propose interventions in order to improve the image of the destination.

Brand communication – Greece

The brand communication of Greece is implemented by the Greek National Tourism Organization[910](GNTO). Supervised by the Hellenic Ministry of Tourism, its main mission is the organization, development and promotion of tourism in Greece by using all the potential of the country, following tourism promotion activities of Greece domestically and abroad and any other activity deemed necessary to support and promote the Greek tourism product. The key goal is to enhance the value of Greek Tourism, in collaboration with the tourism industry and all stakeholders in order to increase incoming tourism and tourism revenues at the same time. GNTO operates with the aim of developing tourism, the most important pillar of economic and social development of the country, being fully aware that its product is distributed to all social groups in the country and directly affects the quality of life of Greek citizens, responding successfully to a highly competitive international environment. The communication objectives during the last years were the creation of a new strong image for the country that would reflect the new vision of Greek tourism as an authentic, anthropocentric, classical, historical, traditional, and at the same time modern destination that has a variety of natural and cultural resources and therefore constitutes a diversified tourism product. The message was simple and revealed the "unknown" authentic modern Greece, emphasizing its unique historical heritage, the traditional Greek hospitality, as well as the feeling of "belonging" and the unparalleled travel experiences it potentially offers to all visitors throughout the year. Therefore, the Logo of the country was "All time Classic." GNTO is active on social media (Instagram, FB, blog, YouTube channel) where all digital campaigns are promoted. Even during the COVID-19 period the country launched an online platform **Greece From Home**[11] that aimed to promote the country's image to people all over the world and invited them to visit – not by traveling physically, but virtually – straight from the comfort and safety of their homes. It was an initiative taken jointly by the Hellenic Ministry of Tourism, GNTO and Marketing Greece and had three goals: to help people from all over the world stay in touch with the country and its culture; for people (future travelers) to discover destinations and be inspired by the country's beauties; for tourism professionals to upgrade their digital skills. All this while staying safe at home. Further campaigns worth mentioning are the following:

- "Visit Greece" app: First official mobile application, which supported the digital transition, providing personalized information, innovative services, free promotion of companies, information on their offers as well as on health protocols for customer protection. Also, the official promotional website of GNTO – www.visitgreece.gr – had an additional column related to COVID-19, providing travelers with all the necessary information and instructions.
- "Destination Greece Health First": International promotion campaign for the safe opening of Greek Tourism carried out in collaboration between the Hellenic Ministry of Tourism and GNTO. Greece was presented worldwide as one of the few truly safe tourist destinations in Europe and around the world. Through Social Media, on the internet with a specific microsite and in combination with the new digital application "Visit Greece," the initiative on the so-called "critical decision," presented in a live and convincing way to the visitors all the health and safety measures applied to promote a safe journey, as well as to regain the trust of all those who wanted to visit the country, but were hesitant because of concern about health issues related to the disease. The emblematic video of GNTO "Destination Greece Health First" was the central element of this advertising campaign, which included an inspired promotion of the brand "Greece," combined with complacency in the compliance of health and safety protocols.
- The advertising campaign "Greek Summer Feeling" prioritized security, aimed at inspiration and emotion, at the idea, values and experiences related to Greece and Greek culture, and of course the insuperability of "Greek Summer." GNTO collaborated with Bloomberg for the international launch of the new digital campaign based on a website with extensive content on bloomberg.com. The campaign communicated all the unique aspects, feelings, thoughts and values that define Greece as a country and as a culture, promoting the experience of the Greek summer.
- The advertising campaign "Greek 4: For You": The Ministry of Tourism, in collaboration with GNTO launched a campaign in August 2020 (which lasted 6 months) to promote four islands in the North-eastern Aegean: Lesvos, Chios, Samos and Leros. This advertising campaign aimed to reposition the destination brand of the four islands and was designed to achieve two goals: to stimulate

tourist flows to these islands and to improve their wider image abroad. The campaign reminded travelers that each destination has its own unique story.

Conclusion

Greece, during the last few years, has set as a core element of its tourism policy, the planning and establishment of DMMOs. The perception on governance has changed, and the involvement of the local community has been set as a top priority in terms of sustainable development. Worldwide, as the role of DMOs in tourism governance is not that familiar – with the exception of Europe – more policy initiatives should be taken toward this direction. This of course requires the collaboration between the public and private sector and local authorities; a condition not always perceived in a positive way, especially by the public central stakeholders. Therefore, the governance approach should be probably modified. The participatory or bottom-up approach is the exception and not the rule in the process of tourism policy planning. Not many countries are willing to approach the private industry or the local communities and cooperate in terms of mutual understanding and sharing knowledge. This paper mainly illustrates that this lack of coordination and cooperation constitutes one of the inhibitory factors resulting in a unilateral strategy, not having taken into consideration the concerns and the voices of the industry and the local stakeholders. A channel of communication, constant dialogue and open discussion between private and public administration should be further enhanced. A strong network between central and decentralized public administration should be built as well, since the issue of fragmented policy actions is a very usual phenomenon in the political scene. Once this challenge is faced, then DMOs will be enabled to operate according to their mission, as tools of inclusive growth, of sustainable tourism development in all aspects, economic, environmental, but most importantly social.

Notes

1 This operational definition of destination management/marketing organization was elaborated by the Committee on Tourism and Competitiveness (CTC) and it was adopted as a recommendation by the 22nd Session of the General Assembly of the UNWTO held on 11–16 September 2017 in Chengdu, China (A/RES/684(XXII)).
2 Association for the Promotion and Development of Tourism, (APDT) Brasov, Romania. AsociatiapentruPromovareasiDezvoltareaTurismului (APDT) Brasov, Romania (in Romanian). www.brasovtourism.eu/
3 https://about.visitberlin.de/en/about-visitberlin, Visit Berlin (accessed 11 February 2021).
4 www.myhelsinki.fi/en/about (accessed 11 February 2021).
5 www.nbtc.nl/en/home/about-us.htm (accessed 11 February 2021).
6 https://zt.zuerich.com/en/about-zt/the-organization (accessed 11 February 2021).
7 www.destinationmanagementplan.gr/en/#files, Destination Management Plan (accessed 10 February 2021).
8 https://news.gtp.gr/2020/12/07/greek-tourism-ministry-presents-new-bill-to-establish-dmmos/ and https://news.gtp.gr/2020/12/07/greek-tourism-ministry-presents-new-bill-to-establish-dmmos/ (accessed 10 February 2021).

9 www.gnto.gov.gr/el, Greek National Tourism Organization (accessed 10 February 2021).
10 www.visitgreece.gr (accessed 10 February 2021).
11 www.greecefromhome.com/ (accessed 8 February 2021).

References

Association for the Promotion and Development of Tourism, (APDT) Brasov, Romania. *AsociatiapentruPromovareasiDezvoltareaTurismului (APDT) Brasov, Romania (in Romanian)*. www.brasovtourism.eu/ (Accessed 18 January 2021).

Beritelli, P., & Laesser, C. (2018). Destination logo recognition and implications on international destination branding by DMOs: A case for saving money. *Journal of Destination Marketing & Management, November 2016*. https://doi.org/10.1016/j.jdmm.2016.08.010.

Bornhorst, T., Brent Ritchie, J. R., & Sheehan, L. (2010). Determinants of tourism success for DMOs & destinations: An empirical examination of stakeholders' perspectives. *Tourism Management, 31*(5), 572–589. https://doi.org/10.1016/j.tourman.2009.06.008.

Foris, D., Florescu, A., Foris, T., & Barabas, S. (2020). Improving the management of tourist destinations: A new approach to strategic management at the DMO level by integrating lean techniques. *Sustainability, 12*, 10201. https://doi.org/10.3390/su122310201.

Lalicic, L., Huertas, A., Moreno, A., & Jabreel, M. (2018). Emotional brand communication on Facebook and Twitter: Are DMOs successful? *Journal of Destination Marketing & Management, 16*, 100350.

Molinillo, S., Liebana-Cabanillas, F., Anaya-Sanchez, R., & Buhalis, D. (2018). DMO online platforms: Image and intention to visit. *Tourism Management, 65*, 116–130. https://doi.org/10.1016/j.tourman.2017.09.021.

UNWTO. (2019). *Guidelines for institutional strengthening of Destination Management Organizations (DMOs) – Preparing DMOs for new challenges*. UNWTO, Madrid. https://doi.org/10.18111/9789284420841 (Accessed 12 February 2021).

UNWTO. (2007). *A practical guide to tourism destination management*. UNWTO, Madrid. https://doi.org/10.18111/9789284412433 (Accessed 11 February 2021).

Uşaklı, A., Koç, B., & Sönmez, S. (2017). How "social" are destinations? Examining European DMO social media usage. *Journal of Destination Marketing & Management, 6*(2), 136–149. https://doi.org/10.1016/j.jdmm.2017.02.001.

Volgger, M., & Pechlaner, H. (2014). Requirements for destination management organizations in destination governance: Understanding DMO success. *Tourism Management, 41*, 64–75, https://doi.org/10.1016/j.tourman.2013.09.001.

8

FACTOR THAT INFLUENCES CONSUMPTION OF UNHEALTHY STREET FOOD

Norhidayah Azman and Albattat Ahmad

Introduction

Food is widely recognized as a crucial attraction for tourists, especially those who engage in international travel. The growing phenomenon of food tourism has led to the recognition of food as an essential element in the marketing strategies employed by tourism destinations (O'Flynn, 2022; Hashim & Jamaluddin, 2022). Consequently, numerous establishments are introducing or creating innovative gastronomic offerings with the aim of attracting tourists and providing them with diverse culinary encounters (Mariani & Okumus, 2022; Saari, 2022; Parahoo & Singh, 2022). The expenditure on food consumption at the traveler's destination is commonly estimated to account for around 40% of their overall budget (Basnyat & Ho, 2022; Liu et al., 2022). In this case, food possesses the capacity to emerge as a prominent tourist attraction within the destination. The sector of food tourism has experienced rapid growth within the tourism industry due to the increasing significance of food in the development of travel plans, commonly referred to as food destinations (Zeynalli & Rahimli, 2022; Ramírez-Hurtado et al., 2022). Prior studies have indicated that food plays a substantial role in shaping the overall tourist experience (Kivela & Crotts, 2006). Additionally, it has been recognized as a strategic tool employed by destinations to attract tourists (Enright & Newton, 2004; Smith & Costello, 2009; Tikkanen, 2007; Yuksel, 2001). The rise and expansion of street food can be attributed to the notable economic advantages derived from tourist activity, as street food possesses various merits such as affordability, accessibility, and limited nutritional value (Abdullah et al., 2022; Sgroi et al., 2022). Street food is a prevalent aspect of urban existence in numerous nations, encompassing both developing countries and those with established economies (Mariani & Okumus, 2022).

DOI: 10.4324/9781032633374-10

Street food plays a crucial role in the culinary landscape of a nation. Street foods refer to the category of ready-to-eat (RTE) foods and beverages that are both prepared and sold by food vendors, predominantly in streets and other public areas. These food items are intended for consumption at a later time without undergoing any additional processing (Wanjari, 2022; Sousa et al., 2022; Gupta, 2022). The consumption of these food items and beverages plays a pivotal role in the dissemination of culinary preferences among both local residents and tourists on a global scale. Additionally, it is imperative to acknowledge their significant contribution to the preservation of cultural and social customs by means of indigenous gastronomy. The phenomenon of street food holds a particular allure for travelers who are interested in exploring the culinary culture and experiencing distinctive flavors, thereby contributing to the growth and development of a country's tourism sector (Trohina & Brînza, 2022; Adeleke & Owoseni, 2022). In addition to generating revenue for sellers, these entities play a crucial role in facilitating job creation. The appeal of these meals is extensive among vendors and customers due to their taste, convenience, affordability, cultural and social significance, and nutritional composition (Omorodion & Ogunekum, 2022; Giasuddin et al., 2022).

It is a common occurrence for individuals to report substantial weight gain upon their return from vacation. However, despite the widespread notion that holidays have a significant impact on body weight, there is a limited body of research investigating this phenomenon (Rezaeipour, 2021; Zhu et al., 2021; Toubes et al., 2021). The primary factor contributing to weight gain is the unrestricted consumption of unhealthy street foods during vacation periods. The significance of safety and health concerns is heightened due to the role of street food as a defining characteristic of tourist destinations (Rezaeipour, 2021; Zhu et al., 2021; Toubes et al., 2021). The detrimental health effects of street food can be attributed to various factors, including the quality of raw ingredients, food processing techniques, and storage practices (World Health Organisation, 2022; Rakha et al., 2022; Baig et al., 2022). Furthermore, it should be noted that the raw materials employed in the manufacturing process of these items exhibit substandard quality. Moreover, these materials have been subjected to improper storage conditions, resulting in unhygienic conditions over prolonged durations (Huriani et al., 2022; Wachenheim, 2021; Pradeilles et al., 2021; Li et al., 2021). The infrastructure of food carts is characterized by minimal provisions for clean drinking water, restroom facilities, freezing capabilities for ice creams, disinfection protocols, hand washing amenities, and garbage management. Due to the intermittent nature of water supply from faucets, the need for water storage is typically imperative. This particular variety of water is unsuitable for the purposes of hand or dishwashing, cooking, or consumption, and concerns regarding contamination are present. Furthermore, street food is susceptible to exposure from various sources, including insects, rodents, pets, and other animals. Additionally, it is subject to unfavorable environmental conditions such as air pollution (Huriani et al., 2022;

Wachenheim, 2021; Pradeilles et al., 2021; Li et al., 2021). Undoubtedly, street food has garnered significant popularity among tourists and has emerged as a primary attraction in numerous destinations. Despite the widespread awareness among tourists regarding the adverse health effects associated with street foods, the challenge of refraining from consuming such unhealthy options during vacation remains considerable. The present study aims to examine the factors that influence tourists' intention to choose unhealthy street food consumption. Additionally, it seeks to explore the role of mediating variables, namely anticipating guilt and anticipating pleasure, in enhancing the desire to consume unhealthy street food during vacation.

Literature review

Location of unhealthy street food

In general, street food enterprises tend to be characterized by their small scale, necessitating minimal skill sets, basic infrastructure, and modest financial resources (Tuffour et al., 2022; Gaffar et al., 2022). Street food vendors typically conduct their business from strategically chosen locations that offer them a competitive advantage. Commonly, these locations are characterized by their popularity, often found in bustling marketplaces and busy intersections. These organizations meet the criteria for being classified as stationary. An alternative type of ambulatory street vendor is characterized by their practice of relocating between different locations, utilizing either bicycles or walking as means of transportation, all the while engaging in the promotion and sale of their merchandize (Nordhagen et al., 2022; Maglumtong & Fukushima, 2022). The primary benefit of this particular category of mobile street food vendors in comparison with their stationary counterparts lies in the enhanced accessibility they offer to customers. This advantage mitigates the mobile vendor's dependence on geographical location. According to Rakha et al. (2022), aside from proximity, the primary advantage for a street food vendor lies in the aspect of physical accessibility, encompassing factors such as an appropriate location and convenient access for pedestrians or different modes of transportation. The availability of unhealthy street food in the marketplace has an impact on tourists' inclination to consume such food (Adeosun et al., 2022).

H1: convenience location of street food products encourages the intention to consume unhealthy street food products during vacation.

Price of unhealthy street food is low

Moreover, the price factor exerts a substantial influence on the consumer's decision-making process. According to the findings of Onsardi et al. (2021), the pricing of a product exerts a substantial influence on consumers' inclination to engage in its consumption or make a purchase. According to Wachyuni and Wiweka (2022), street food operations possess certain characteristics that contribute to their ability to maintain low prices for their products. These characteristics include simplified food preparation methods, a relatively modest initial investment, reduced labor demands, and lower overhead costs. The pricing of unhealthy street meals may have an influence on the inclination of tourists to acquire such foods, as individuals tend to place importance on price when assessing the value of street food (Sgroi et al., 2022). This approach aligns seamlessly with street food, which is commonly characterized by its affordability. The benefits associated with street food meals encompass their affordability and immediate availability, factors that tend to foster the consumption of unhealthy street food among tourists during their vacations.

H2: Low price of unhealthy street food products influences the intention to consume unhealthy street food products during vacation

Perceive healthiness

The inclusion of health-related information, such as ingredients, reduced calories, cooking methods, and health claims, in unhealthy food promotions is infrequent, resulting in a limited reduction of customers' perception of the unhealthiness of the advertised food item. The concept of perceived healthiness refers to the cognitive response elicited by customers when assessing the degree of healthfulness of a food product, taking into account both objective and subjective information (Finlay et al., 2022; Medina et al., 2022; Carters-White et al., 2022). The perception of healthiness is an attribute associated with the quality of food and serves as an indicator of satisfaction with a particular food item (Kombanda et al., 2022). Furthermore, alongside cognitive responses, affective responses also play a significant role in shaping consumers' decision-making processes. Extensive theoretical analyses have consistently shown that the affective component of attitudes is subsequent to cognitive processes (Edwards, 1990). To clarify, affective responses are derived from evaluations of beliefs that are based on expectations (Fishbein & Ajzen, 1977). According to Berkowitz (1993), the emotional-cognitive model of decision-making posits that affective reactions are influenced by cognitive processing, whereby cognitive processes can either enhance or diminish the strength of these reactions. Thus, affective reactions are believed to arise as a result of cognitive processes.

Consumers' perceptions of the healthiness of advertised unhealthy street food are influenced by a cognitive process that takes into account both intrinsic factors (such as past experience, motivation, and personal beliefs) and extrinsic factors (including advertised claims, calorie content, and nutrient information) (Iancu Iancu, 2022). The impact of individuals' cognitive perception of healthiness on their emotional response to anticipated feelings is a significant factor in the domain of healthy food consumption (Finlay et al., 2022; Medina et al., 2022; Carters-White et al., 2022). Consumers have the ability to deduce anticipated emotional reactions by assessing the nutritional worth of a food item and its subsequent consequences. According to Winkler et al. (2022), in cases where consumers are unable to make accurate assessments regarding the nutritional value of a food product based solely on the information provided on the nutritional label, they experience a reduced sense of shame. In essence, mitigating the most detrimental perceptions could potentially alleviate the subjective experience of being in error or the fear of being judged negatively by others. According to O'Hagan (202), the ability to accurately assess the nutritional value of food instills a sense of ethical responsibility in consumers. Based on these assumptions, the current study posits that there is a positive correlation between an increase in perceived unhealthiness and the anticipated reduction in guilt and enhancement of pleasure.

H3: perceived healthiness of the unhealthy street foods products positively influences to reduce the guilt toward to consume unhealthy street food products during vacation.

H4: perceived healthiness of the unhealthy street food products positively influences anticipated pleasure toward the intention to consume unhealthy street food products during vacation.

The effect of anticipated guilt and pleasure

The assessment of the nutritional value of a street meal is influenced by individuals' cognitive reactions, which are shaped by their objective and subjective understanding. The perception of healthiness is commonly regarded as an attribute of street food quality and a potential indicator of food consumption (Mohamad et al., 2022; Savelli et al., 2022). In addition to the cognitive response, consumers' decision-making processes are also influenced by emotional responses. Multiple theoretical analyses have demonstrated that the affective component of attitudes is postcognitive, as evidenced by studies conducted by Ortony et al. (2022) and Gaekwad et al. (2022). To clarify, affective responses are derived from evaluations of beliefs that are based on expectations (Fishbein & Ajzen, 1975). Berkowitz (1993) proposed the emotional-cognitive model of decision-making, which posits that affective reactions are influenced by cognitive processing, leading to the emergence of affective reactions as a result of cognitive processes. The perception of the healthiness of street food products that are promoted to consumers in the absence of information regarding unhealthy food is influenced by a cognitive process that takes into account both internal factors (such as prior experience, motivation, and personal beliefs) and external factors (such as advertising, calorie content, and nutrient information) (Sousa, 2022). The cognitive evaluation of perceived healthiness plays a role in shaping the affective response of anticipated emotions within the context of consuming healthy food. Tourists have the potential to deduce likely emotional consequences by evaluating the perceived quality and subsequent physiological effects of the local cuisine. When consumers perceive the nutritional value of advertised unhealthy street foods in a positive manner, they experience a reduction in feelings of guilt. In alternative terms, heightened states of well-being have the potential to mitigate sensations of personal misalignment or the perception of such misalignment by external individuals. Likewise, when consumers perceive that the unhealthy fast food products being promoted possess greater nutritional value, their expected level of satisfaction escalates. According to Kiss et al. (2022), individuals experience a sense of moral satisfaction when they are able to accurately assess the nutritional value of street food. Based on the aforementioned assumptions, the present study posits that an increase in perceived healthiness is inversely correlated with anticipated guilt, while positively correlated with anticipated pleasure.

The feedback system plays a crucial role as a fundamental component of anticipated emotions within the context of decision-making (Meijerink, 2022). As per the concept of feedback, emotions serve as an intrinsic mechanism for incentivizing or penalizing behavior (Meijerink, 2022). According to Gollwitzer (1999), individuals' behavior is often influenced by pre-existing conditional statements or personal aspirations, rather than being solely determined by the immediate behavioral options at hand. According to Gollwitzer (1999), emotions play a crucial role in the process of if-then conditional rules by offering timely feedback on recent behaviors and directing individuals toward appropriate actions. The persuasive nature of emotional input stems from the inherent inability of individuals to exert direct control over their emotions, in contrast to their capacity to directly control their cognitive processes (Gollwitzer, 1999). For example, individuals have a tendency to conveniently disregard or intentionally ignore the unfavorable outcomes of their actions, yet the elimination of the adverse emotions evoked by such behaviors proves to be challenging. According to Polman et al. (2022), anticipated emotions play a crucial role in influencing individuals' future actions, as they prompt individuals to contemplate potential emotional consequences and are regarded as dependable indicators of future behavior. Positive anticipated emotions often serve to validate an activity, as they signify that the outcome is likely to be favorable and therefore serve as a motivating factor for engaging in the behavior. In contrast, negative emotions serve as indicators of potential ineffectiveness in an individual's actions, suggesting the need for behavioral adjustments (Meijerink, 2022).

In the context of decision-making regarding unhealthy food choices, individuals tend to anticipate experiencing minimal guilt and heightened pleasure. This anticipation is influenced by cognitive responses, specifically the perception of certain food options as being particularly unhealthy, as well as the outcomes of recent behaviors. For instance, individuals may feel little guilt when consuming unhealthy street food or derive pleasure from maintaining an unhealthy diet while on vacation. For instance, if an individual expects to experience minimal guilt when consuming unhealthy food products while on vacation, it is likely to enhance their intention to purchase such items and potentially lead them to modify or substitute their behaviors. This is because the anticipation of low guilt does not serve as a reminder of the ethical implications of their actions or provide guidance on how to avoid negative consequences (Meijerink, 2022).

The desire for anticipated enjoyment motivates individuals to acquire and consume unhealthy food items while on vacation, as it leads them to believe that such actions are justified based on prior positive experiences or alignment with personal objectives. Furthermore, previous studies utilizing the Mehrabian–Russell (M-R) model have demonstrated the significant influence of positive and negative emotions on various behavioral intentions within the context of restaurants. These intentions include repeat purchases, recommendations, and positive word of mouth (Isaacs et al., 2022; Singh & Sharma, 2022). Based on the aforementioned assumptions, the current study has formulated a hypothesis.

H5: Low guilt toward the unhealthy food products consumption positively influences behavioural intentions to consume unhealthy street food products products during vacation.

H6: Anticipated pleasure toward the unhealthy food products consumption positively influences behavioural intention to consume unhealthy street food products during vacation.

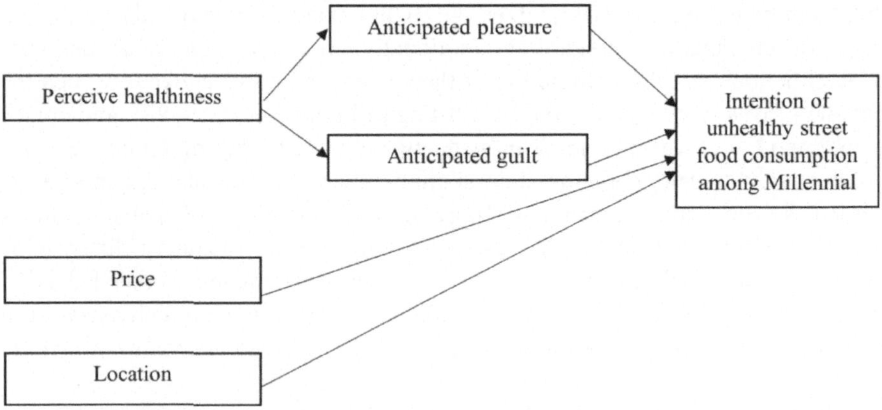

FIGURE 8.1 Conceptual framework

H7: Low guilt toward unhealthy street food products consumption mediates the relationship between perceived healthiness and intention to consume unhealthy food products consumption during vacation

H8: Anticipate pleasure toward unhealthy street food products consumption mediates the relationship between perceive healthiness and intention to consume unhealthy food products consumption during vacation

Research methodology

The primary goal of this study is to investigate factors that influence the intention to consume unhealthy street food during vacation and the influence of the mediation effect of anticipated guilt and pleasure. To fulfill this study objective, a self-administered questionnaire has been used. The survey questionnaire consisted of three parts which demographic profile for Part A and Part B consisted of questions for location, price, and perceive healthiness. Part C consists of question for the mediation variable which is anticipated guilt and pleasure and finally Part D, consist of a questionnaire on the intention to choose unhealthy street food; 650 tourists have volunteered to participate in this study. Questionnaires have been distributed through Google Forms and analyzed by using Smart Pls.

Finding

Internal consistency, as determined by the composite reliability (CR) score, was used in determining the validity and reliability of this study. Three indications are used to assess the dependability of each item. It must approach above 0.70 for composite reliability (Fornell & Larcker, 1981), above 0.70 for Cronbach's alpha (CA) (Fornell & Larcker, 1981; Nunnally & Bernstein, 1994), and above 0.50 for average variance extracted (AVE) (Fornell & Larcker, 1981). From Table 8.1, the result indicates value for CR is above 0.70 with a range of 0.771 to 0.942 which is accepted according to Fornell and Larcker (1981). For CA value, the score presented

TABLE 8.1 CA, CR and AVE values

	Cronbach's Alpha	*rho_A*	*Composite Reliability*	*Average Variance Extracted (AVE)*
GUILTY	0.929	0.954	0.942	0.618
HAPPY EMOTION	0.851	0.906	0.888	0.616
INTENTION	0.870	0.879	0.880	0.717
PERCEIVE HEALTHINESS	0.876	0.898	0.895	0.665
PLEASURE	0.881	0.914	0.910	0.510
TASTES	0.759	0.863	0.771	0.626

in Table 8.1 reported all values exceed 0.70 with the range 0.759 to 0.929. This indicates all the value is accepted. Finally, for AVE result, the score presented shows more than 0.50 with the range of 0.510 to 0.626 (Fornell & Larcker, 1981; Nunnally & Bernstein, 1994) as presented in Table 8.1.

Discriminant validity is the extent to which experimentally distinct constructs differ from one another in a meaningful way. In addition, it estimates the degree of difference between the overlapping conceptions. One can evaluate the discriminant validity using cross-loading of indicator, the Fornell & Larcker criteria, and the heterotrait–monotrait (HTMT) correlation ratio. By assessing cross-loading, the factor loading indicators on the assigned construct must be more than the loading of the other constructs if the factor loading cut-off value is greater than 0.70 (Hair et al., 2011, 2017). The evaluation of discriminant validity using the Fornell–Lacker criterion is the second criterion (Fornell et al., 1994). This method compares the square root of the average variance extracted (AVE) to the correlation between latent constructs (Hair et al., 2017). A latent construct must better explain the variation of its own indicator than the variance of other latent constructs. Consequently, the square root of each construct's AVE should be greater than its correlations with other latent constructs (Hair et al., 2017). In this study, Table 8.3 demonstrates that the square roots of AVE for each construct are greater than the construct's strongest correlation with other latent variables in the model, when compared to the AVE result reported in Table 8.2.

The correlation ratio between heterotraits and monotraits is the second measure of discriminant validity (HTMT). Henseler et al. (2015) determined, using Monte Carlo simulation research, that HTMT achieves higher specificity and sensitivity rates (97 to 99 percent) than the cross-loadings criteria (0.00 percent) and Fornell–Lacker technique. When HTMT levels are close to 1, there is a lack of discriminant validity. To use the HTMT as a criterion, it must be measured against a preset threshold. If the HTMT score surpasses this threshold, the test's discriminant validity may be questioned. Some authors advocate a threshold of 0.85 (Kline, 2011). In addition, Gold et al. (2001) argued against it and proposed an alternative value of

TABLE 8.2 Discriminant validity

	GUILTY	*LOCATION*	*INTENTION*	*PERCEIVE HEALTHINESS*	*PLEASURE*	*PRICE*
GUILTY	0.869					
LOCATION	0.850	0.815				
INTENTION	0.651	0.568	0.846			
PERCEIVE HEALTHINESS	0.754	0.758	0.421	0.882		
PLEASURE	0.668	0.742	0.257	0.640	0.818	
PRICE	0.721	0.617	0.509	0.671	0.615	0.853

TABLE 8.3 Heterotraits and monotraits (HTMT)

	GUILTY	*LOCATION*	*INTENTION*	*PERCEIVE HEALTHINESS*	*PLEASURE*	*PRICE*
GUILT						
LOCATION	0.719					
INTENTION	0.614	0.739				
PERCEIVE HEALTHINESS	0.761	0.802	0.803			
PLEASURE	0.776	0.807	0.801	0.709		
PRICE	0.799	0.739	0.626	0.841	0.790	

TABLE 8.4 R^2 value

	R Square	*R Square Adjusted*
GUILTY	0.568	0.567
INTENTION	0.524	0.519
PLEASURE	0.409	0.408

0.90. According to the score in Table 8.3, the HTMT value surpasses the threshold value proposed by Kline (2011), indicating that the discriminant validity of the current study is less than 0.85.

The coefficient of determination (R^2) measures the amount of variance explained by endogenous variables (Hair et al., 2011). R^2 is used to determine a structural model's explanatory power (Ringle et al., 2014). R^2 must be good, where values of 0.25, 0.50, and 0.75 are deemed weak, medium, and considerable, respectively, for target constructions (Hock & Ringle, 2010). According to Table 8.4, perceive healthiness to anticipate guilt score 0.568 which consider medium explanatory power which indicate perceive healthiness explain 56.8% of anticipated guilt. For anticipate pleasure, the R2 value score 0.409 which consider weak explanatory power and location, price, perceive healthiness, anticipate pleasure, and anticipate guilt explain 0.524 of intention to consume unhealthy street food.

The results of hypothesis testing for the direct relationship presented in Table 8.5 indicate that all relationships are statistically significant and supported. Hypothesis 1 suggests that the presence of conveniently located street food establishments influences individuals' inclination to consume unhealthy street food while on vacation, as indicated by a statistically significant regression coefficient ($\beta = 0.214$, $t = 3.349$, $p < 0.005$). Based on the correlation score, it can be inferred that Hypothesis 1 exhibits statistical significance. Hypothesis 2 posits that a decrease in the price of unhealthy street food would have an impact on individuals' inclination to consume such food while on vacation, as indicated by the coefficient

TABLE 8.5 Hypothesis testing – direct relationship

	Original Sample (O)	*Sample Mean (M)*	*Standard Deviation (STDEV)*	*T Statistics (|O/STDEV|)*	*P Values*
LOCATION →INTENTION	0.214	0.226	0.064	3.349	0.001
PRICE →INTENTION	0.354	0.353	0.077	4.624	0.000
PERCEIVE HEALTHINESS → GUILT	0.754	0.757	0.016	47.388	0.000
PERCEIVE HEALTHINESS → PLEASURE	0.640	0.642	0.042	15.376	0.000
GUILT →INTENTION	0.520	0.511	0.089	5.844	0.000
PLESURE →INTENTION	0.486	0.485	0.058	8.376	0.000

score ($\beta = 0.354$, $t = 4.624$, $p < 0.005$). Hypothesis 3 posits that the perception of healthiness associated with unhealthy street foods has a positive impact on reducing feelings of guilt toward consuming such products while on vacation. Based on the findings, the correlation analysis reveals a significant positive relationship between the variables, as evidenced by the beta coefficient ($\beta = 0.754$, $t = 47.388$, $p < 0.005$). The significance of urge Hypothesis 3 is noteworthy. Hypothesis 4 posits that the perception of healthiness of unhealthy street food has a positive effect on the anticipated pleasure associated with the intention to consume such products during vacation. The correlation analysis, as depicted in Table 8.5, reveals a significant relationship with a beta coefficient of 0.640, a t-value of 15.376, and a p-value less than 0.005. There is a compelling argument for the significance of hypothesis 4. In relation to Hypothesis 5, the correlation score is statistically significant ($\beta = 0.520$, $t = 5.844$, $p < 0.005$). This finding suggests that individuals who experience low levels of guilt regarding their consumption of unhealthy food are more likely to have intentions to consume unhealthy street food while on vacation. Finally, Hypothesis 6 is proposed. The expectation of experiencing pleasure from consuming unhealthy food has a positive impact on the intention to consume unhealthy street food while on vacation. The obtained correlation score ($\beta = 0.486$, $t = 8.376$, $p < 0.005$) provides evidence to support the significance of hypothesis 6.

The mediation results for the perception of guilt and pleasure are presented in Table 8.6. Hypothesis 7 posits that a lower sense of guilt associated with consuming

TABLE 8.6 Hypothesis testing – mediation effect

	Original Sample (O)	*Sample Mean (M)*	*Standard Deviation (STDEV)*	*T Statistics (|O/STDEV|)*	*P Values*
PERCEIVE HEALTHINESS →GUILT →INTENTION	0.392	0.386	0.071	5.547	0.000
PERCEIVE HEALTHINESS →PLEASURE →INTENTION	0.311	−0.312	0.046	6.682	0.000

unhealthy street food acts as a mediator between perceived healthiness and the intention to consume such food during vacations. Hypothesis 8 suggests that the anticipation of pleasure derived from consuming unhealthy street food mediates the relationship between perceived healthiness and the intention to consume such food during vacations. The correlation scores for both hypotheses indicate significant relationships, with ($\beta = 0.392$, $t = 5.547$, $p < 0.005$) for the first hypothesis and ($\beta = 0.311$, $t = 6.628$, $p < 0.005$) for the second hypothesis. This finding suggests that both hypotheses are supported.

Discussion

Unhealthy street food is typically found in public areas and makes use of public facilities. Street cuisine, also known as street food, is typically observed in public spaces such as streets, markets, or fairs. The culinary offering in question is frequently dispensed from a mobile food booth, food cart, or food truck, with the primary aim of facilitating expeditious consumption (Nordhagen et al., 2022; Maglumtong & Fukushima, 2022). According to Rakha et al. (2022), the convenient availability of unhealthy street food in close proximity played a role in influencing tourists to opt for such food options during their vacations. The presence of strategic locations in tourist destinations has been found to influence the inclination toward consuming unhealthy street food while on vacation (Tuffour et al., 2022; Gaffar et al., 2022). The cost of unhealthy street food is significantly lower in comparison to that of healthy food. This is due to the fact that street food vendors tend to operate in tourist-heavy areas, resulting in a similarity in the types of food offered. Consequently, pricing becomes a key differentiating factor among vendors, serving as a promotional strategy (Wachyuni & Wiweka, 2022). The affordability of street food often serves as a compelling factor in enticing tourists to make purchases. Notably, the pricing of street food, whether it is deemed nutritious or detrimental to one's health, has proven to be an effective marketing strategy for the promotion of unhealthy street food (Sgroi et al., 2022; Seo & Lee, 2021).

Engaging in the consumption of local street food is a prevalent activity among tourists during their vacations, as it allows them to experience the distinctiveness and cultural variations of the destination (Finlay et al., 2022; Medina et al., 2022; Carters-White et al., 2022). Furthermore, food can serve as a means to convey the cultural variations that exist globally, with street food being particularly effective in this regard (Iancu Iancu, 2022). Street food is commonly prepared using locally sourced ingredients and by local individuals who possess a deep understanding of the specific types of local cuisine they aim to offer to tourists. The lack of nutritional information provided by street food vendors tends to enhance tourists' perception of the healthiness of their food (Winkler et al., 2022; O'Hagan, 2022). Street food is commonly perceived as a trustworthy option for consumption due to its characteristic of being freshly prepared in the presence of the consumer, thereby instilling confidence in the food's healthiness. The lack of awareness regarding local unhealthy ingredients and food preparation practices leads tourists to mistakenly perceive the food they consume as healthy. Furthermore, this phenomenon also engenders a sense of anticipation and ambivalence among tourists (Kombanda et al., 2022).

Conclusion

The acknowledgment of the escalating importance of the consumption of unhealthy street food among tourists has heightened the necessity to introduce healthy street food options in the market. The study revealed significant associations between the consumption of unhealthy street food among tourists in the market and factors such as location, price, perceived healthiness, as well as the mediating effects of anticipated guilt and pleasure. It is imperative for the government to recognize the necessity of implementing a street food management program in order to incentivize street food vendors to enhance the nutritional quality of their offerings. This measure is crucial to ensure that the street food provided to tourists is healthy and adheres to appropriate health standards (Omorodion & Ogunekum, 2022). Furthermore, it is imperative for the government to establish an information center that disseminates comprehensive details about the nutritional content of local street food to tourists. This measure is crucial in ensuring that tourists are well-informed about the specific types of street food available to them. The aforementioned studies (Ortony et al., 2022; Gaekwad et al., 2022) suggest that the provision of healthy street food options can serve as a motivating factor for tourists when making their dietary choices during vacation. According to the study conducted by Ma et al. (2019), the presence of the information center will also serve as a means of notifying local street food vendors about the importance of offering nutritious street food options to tourists. This is in response to the government's efforts to raise awareness about the potential health risks associated with consuming unhealthy street food (Hasan, 2022; Ozcelik & Akova, 2021). Furthermore, it is imperative for street vendors to assume the responsibility of cultivating and executing a strong sense of integrity within themselves. This is crucial in order to guarantee the production and sale of wholesome street food products to their customers.

References

Abdullah, T., Carr, N., & Lee, C. (2022). The empowerment of street food vendors, a marginalised community within the hospitality industry. *Tourism Recreation Research*, 1–14.

Adeleke, O. A., & Owoseni, A. A. (2022). Antibiotic resistance and presence of plasmids in bacteria isolated from cooked street foods. *International Journal of Environmental Studies*, 1–8.

Adeosun, K. P., Greene, M., & Oosterveer, P. (2022). Urban daily lives and out-of-home food consumption among the urban poor in Nigeria: A practice-based approach. *Sustainable Production and Consumption*, *32*, 479–491.

Baig, J. A., Chandio, I. G., Kazi, T. G., Afridi, H. I., Akhtar, K., Junaid, M., . . . Malghani, N. A. (2022). Risk assessment of macronutrients and minerals by processed, street, and restaurant traditional Pakistani foods: A case study. *Biological Trace Element Research*, 1–14.

Basnyat, S., & Ho, I. T. E. (2022). Food and tourist experiences: Insights from Macau. *Journal of Global Scholars of Marketing Science*, *32*(1), 97–113.

Berkowitz, L. (1993). Pain and aggression: Some findings and implications. *Motivation and Emotion*, *17*, 277–293.

Carters-White, L. E., Patterson, C., Nimegeer, A., Hilton, S., & Chambers, S. (2022). Newspaper framing of food and beverage corporations' sponsorship of sport: A content analysis. *BMC Public Health*, *22*(1), 1–11.

Enright, M. J., & Newton, J. (2004). Tourism destination competitiveness: A quantitative approach. *Tourism Management*, *25*(6), 777–788.

Edwards, K. (1990). The interplay of affect and cognition in attitude formation and change. *Journal of Personality and Social Psychology*, *59*(2), 202.

Finlay, A., Robinson, E., Jones, A., Maden, M., Cerny, C., Muc, M., . . . Boyland, E. (2022). A scoping review of outdoor food marketing: Exposure, power and impacts on eating behaviour and health. *BMC Public Health*, *22*(1), 1–48.

Fishbein, M., & Ajzen, I. (1977). Belief, attitude, intention, and behavior: An introduction to theory and research.

Fornell, C., Cha, J., & Bagozzi, R. P. (1994). *Advanced methods of marketing research*. Blackwell Business.

Fornell, C., & Larcker, D. F. (1981). Evaluating structural equation models with unobservable variables and measurement error. *Journal of Marketing Research*, *18*(1), 39–50.

Gaekwad, J. S., Moslehian, A. S., Roös, P. B., & Walker, A. (2022). A meta-analysis of emotional evidence for the biophilia hypothesis and implications for biophilic design. *Frontiers in Psychology*, *13*.

Gaffar, V., Tjahjono, B., Abdullah, T., Sari, M., & Rofaida, R. (2022). Unfolding the impacts of a prolonged COVID-19 pandemic on the Sustainability of Culinary tourism: Some insights from micro and small street food vendors. *Sustainability*, *14*(1), 497.

Giasuddin, A. S. M., Ahmed, A., & Rahman, S. M. (2022). Implications of safe street foods in Dhaka city: An opinion. *European Journal of Nutrition & Food Safety*, 58–63.

Gold, A. H., Malhotra, A., & Segars, A. H. (2001). Knowledge management: An organizational capabilities perspective. *Journal of Management Information Systems*, *18*(1), 185–214.

Gollwitzer, P. M. (1999). Implementation intentions: Strong effects of simple plans. *American Psychologist*, *54*(7), 493.

Gupta, V. (2022). Street food. In *Encyclopedia of tourism management and marketing* (pp. 272–274). Edward Elgar Publishing.

Hair, J. F., Ringle, C. M., & Sarstedt, M. (2011). The use of partial least squares (PLS) to address marketing management topics. *Journal of Marketing Theory and Practice*, *19*(2), 135–138.

Hair, Jr., J. F., Sarstedt, M., Ringle, C. M., & Gudergan, S. P. (2017). *Advanced issues in partial least squares structural equation modeling*. Sage Publications.

Hasan, R. (2022). *Entrepreneurial behavior in informal economy: A case study on street food entrepreneurship in Dhaka, Bangladesh* (Doctoral dissertation, University of Essex and University of Suffolk).

Hashim, R., & Jamaluddin, M. R. (2022). Co-creation of destination image and food tourism of the emerging economies. In *Gastronomy, hospitality, and the future of the restaurant industry: Post-COVID-19 perspectives* (pp. 244–257). IGI Global.

Henseler, J., Ringle, C. M., & Sarstedt, M. (2015). A new criterion for assessing discriminant validity in variance-based structural equation modeling. *Journal of the Academy of Marketing Science*, *43*, 115–135.

Hock, M., & Ringle, C. M. (2010). Local strategic networks in the software industry: An empirical analysis of the value continuum. *International Journal of Knowledge Management Studies*, *4*(2), 132–151.

Huriani, Y., Haryanti, E., Zulaiha, E., & Haq, M. Z. (2022). Women religious congregation as driving force behind alleviation of urban poor nutrition. *Cogent Social Sciences*, *8*(1), 2113599.

Iancu Iancu, J. (2022). *Food tourism promotion of Norway – Unique, weird and wonderful.*

Isaacs, A., Halligan, J., Neve, K., & Hawkes, C. (2022). From healthy food environments to healthy wellbeing environments: Policy insights from a focused ethnography with low-income parents' in England. *Health & Place*, *77*, 102862.

Kiss, M., Czine, P., Balogh, P., & Szakály, Z. (2022). The connection between manufacturer and private label brands and brand loyalty in chocolate bar buying decisions – A hybrid choice approach. *Appetite*, *177*, 106145.

Kivela, J., & Crotts, J. C. (2006). Tourism and gastronomy: Gastronomy's influence on how tourists experience a destination. *Journal of Hospitality & Tourism Research*, *30*(3), 354–377.

Kline, R. B. (2023). *Principles and practice of structural equation modeling*. Guilford Publications.

Kombanda, K. T., Margerison, C., Booth, A., & Worsley, A. (2022). How young adults in Australia classify foods as healthy and unhealthy. *Appetite*, *175*, 106060.

Li, J., Ma, W., & Yu, T. (2021). *Rural food environment, caregiver influence and school-aged children's unhealthy food choice: New evidence from Rural China.*

Liu, T., Juvan, E., Qiu, H., & Dolnicar, S. (2022). Context-and culture-dependent behaviors for the greater good: A comparative analysis of plate waste generation. *Journal of Sustainable Tourism*, *30*(6), 1200–1218.

Ma, L., Chen, H., Yan, H., Wu, L., & Zhang, W. (2019). Food safety knowledge, attitudes, and behavior of street food vendors and consumers in Handan, a third tier city in China. *BMC Public Health*, *19*(1), 1–13.

Maglumtong, M., & Fukushima, S. (2022). Inclusive development for street food vendors in modernizing Bangkok, Thailand: An evaluation of the street food program of the government savings bank. *Urban and Regional Planning Review*, *9*, 42–62.

Mariani, M., & Okumus, B. (2022). Features, drivers, and outcomes of food tourism. *British Food Journal*, *124*(2), 401–405.

Medina, C., Piña-Pozas, M., Aburto, T. C., Chavira, J., López, U., Moreno, M., . . . Barquera, S. (2022). Systematic literature review of instruments that measure the healthfulness of food and beverages sold in informal food outlets. *International Journal of Behavioral Nutrition and Physical Activity*, *19*(1), 1–14.

Meijerink, J. J. (2022). *What is really your excuse? Exploring how Millennials cope with attitudinal-behavioral gaps in the domains of health and sustainability* (Bachelor's thesis, University of Twente).

Mohamad, N., Palan, D. S., Roslan, M. A., & Nasron, N. A. (2022). Predictors of behavioral intention among tourist: The case of revisiting street food spots in Penang, Malaysia. *Journal of Foodservice Business Research*, *25*(4), 475–497.

Nordhagen, S., Fofana, M. L., Diallo, S., Songbono, J. L., Stokes-Walters, R., Zhang, L. X., . . . Winch, P. J. (2022). Between the city and the farm: Food environments in artisanal mining communities in Upper Guinea. *Public Health Nutrition*, *25*(2), 368–380.

Nunnally, J. C., & Bernstein, I. H. (1994). *Psychometric theory New York*. McGraw-Hill.

O'Flynn, R. (2022). *Back to the future: The artisan food producer in Ireland's food tourism proposition*.

O'Hagan, L. A. (2022). "Foodstagramming" in early 20th-century postcards: A transhistorical perspective. *Visual Communication*, 14703572221096715.

Omorodion, N. J., & Ogunekum, A. (2022). Microbial and physicochemical screening of ready to eat street foods. *Asian Journal of Basic Science & Research*, *4*(1), 24–35.

Onsardi, O., Wulandari, K., Finthariasari, M., & Yulinda, A. T. (2021). Impact of service marketing on student decisions. *JBMP (Jurnal Bisnis, Manajemen dan Perbankan)*, *7*(2), 234–254.

Ortony, A., Clore, G. L., & Collins, A. (2022). *The cognitive structure of emotions*. Cambridge University Press.

Ozcelik, A., & Akova, O. (2021). The impact of street food experience on behavioural intention. *British Food Journal*, *123*(12), 4175–4193.

Parahoo, S. K., & Singh, A. S. (2022). Culinary tourism motivations. In *Encyclopedia of tourism management and marketing*. Edward Elgar Publishing.

Polman, E., Ruttan, R. L., & Peck, J. (2022). Using curiosity to incentivize the choice of "should" options. *Organizational Behavior and Human Decision Processes*, *173*, 104192.

Pradeilles, R., Irache, A., Wanjohi, M. N., Holdsworth, M., Laar, A., Zotor, F., . . . Griffiths, P. (2021). Urban physical food environments drive dietary behaviours in Ghana and Kenya: A photovoice study. *Health & Place*, *71*, 102647.

Rakha, A., Fatima, M., Bano, Y., Khan, M. A., Chaudhary, N., & Aadil, R. M. (2022). Safety and quality perspective of street vended foods in developing countries. *Food Control*, 109001.

Ramírez-Hurtado, J. M., Oliva-López, F., & Berbel-Pineda, J. M. (2022). Antecedents and outcomes of memorable tourism experiences in tourism recreation: The case of motorcycle tourism. *Tourism Recreation Research*, 1–14.

Rezaeipour, M. (2021). COVID-19-related weight gain in school-aged children. *International Journal of Endocrinology and Metabolism*, *19*(1).

Ringle, C. M., Sarstedt, M., & Schlittgen, R. (2014). Genetic algorithm segmentation in partial least squares structural equation modeling. *OR Spectrum*, *36*, 251–276.

Smith, S., & Costello, C. (2009). Culinary tourism: Satisfaction with a culinary event utilizing importance-performance grid analysis. *Journal of Vacation Marketing, 15*(2), 99–110.

Saari, S. (2022). In search for new urban tourism niche: Could European cities be destinations for urban wellness providing food for body, mind and spirit? *Tourism and Hospitality Research*, 14673584221086888.

Savelli, E., Gregory-Smith, D., Murmura, F., & Pencarelli, T. (2022). How to communicate typical – Local foods to improve food tourism attractiveness. *Psychology & Marketing, 39*(7), 1350–1369.

Seo, K. H., & Lee, J. H. (2021). Understanding risk perception toward food safety in street food: The relationships among service quality, values, and repurchase intention. *International Journal of Environmental Research and Public Health, 18*(13), 6826.

Sgroi, F., Modica, F., & Fusté-Forné, F. (2022). Street food in Palermo: Traditions and market perspectives. *International Journal of Gastronomy and Food Science, 27*, 100482.

Singh, G., & Sharma, S. (2022). Obese customers' fitness goal disclosure on social media: Exploring weight-loss image sharing on emotions and healthy lifestyle aspirations. *European Journal of Marketing, 56*(11), 3138–3167.

Sousa, S., Albuquerque, G., Gelormini, M., Casal, S., Pinho, O., Damasceno, A., . . . Padrão, P. (2022). Nutritional content of the street food purchased in Chişinău, Moldova: Opportunity for policy action. *International Journal of Gastronomy and Food Science, 27*, 100456.

Tikkanen, I. (2007). Maslow's hierarchy and food tourism in Finland: Five cases. *British Food Journal, 109*(9), 721–734.

Toubes, D. R., Araújo Vila, N., & Fraiz Brea, J. A. (2021). Changes in consumption patterns and tourist promotion after the COVID-19 pandemic. *Journal of Theoretical and Applied Electronic Commerce Research, 16*(5), 1332–1352.

Trohina, A., & Brînza, R.-C. (2022). *Risk of developing the bakery street food network in mun: Chisinau on public health.* Conferinţa tehnico-ştiinţifică a studenţilor, masteranzilor şi doctoranzilor, 29–31 martie 2022, Chişinău. Chişinău, Republica Moldova: Tehnica-UTM, 2022, Vol. 1 (pp. 581–585). ISBN 978-9975-45-828-3.

Tuffour, J. K., Oppong, M., Nyanyofio, J. G. T., Abukari, M. F., Addo, M. N., & Brako, D. (2022). Micro-entrepreneurship: Examining the factors affecting the success of women street food vendors. *Global Business Review*, 09721509211072380.

Wachenheim, C. (2021). Situated learning: Food safety among Chinese food vendors. *Journal of Food Science Education, 20*(4), 155–165.

Wachyuni, S. S., & Wiweka, K. (2022). Tourists' perceptions of local street food: Angkringan, Yogyakarta's culinary icon. *International Journal of Travel, Hospitality and Events, 1*(3), 239–247.

Wanjari, S. T. (2022). Microbial profiling of street foods available at different locations. *Journal of Coastal Life Medicine, 10*, 11–16.

Winkler, M. R., Lenk, K., Erickson, D. J., & Laska, M. N. (2022). Retailer marketing strategies and customer purchasing of sweetened beverages in convenience stores. *Journal of the Academy of Nutrition and Dietetics, 122*(11), 2050–2059.

World Health Organization. (2022). *WHO/FAO inter-regional meeting to promote healthy diets through the informal food sector in Asia: 20–22 August 2019, Bangkok, Thailand* (No. SEA-NUT-201). World Health Organization. Regional Office for South-East Asia.

Yuksel, A. (2001). Managing customer satisfaction and retention: A case of tourist destinations, Turkey. *Journal of Vacation Marketing*, *7*(2), 153–168.
Zeynalli, L., & Rahimli, E. (2022). The role of human capital in increasing tourism potential in a post-conflict situation. *Future Human Image*, *17*, 101–10.
Zhu, Q., Li, M., Ji, Y., Shi, Y., Zhou, J., Li, Q., . . . Zhuang, X. (2021). "Stay-at-home" lifestyle effect on weight gain during the COVID-19 outbreak confinement in China. *International Journal of Environmental Research and Public Health*, *18*(4), 1813.

9

SUSTAINABLE TOURISM IN THE POST-COVID ERA

The Case of Athens destination

Panoraia Poulaki, Antonios Kritikos, Nikolaos Vasilakis and Marco Valeri

Introduction

The tourism industry is considered unsustainable and vulnerable to risks that can cause crises or disasters. When choosing a tourism destination, a large percentage relies on safety, so potential risks negatively affect a visitor's decision.

In addition to the enormous human suffering and tens of thousands of deaths caused by the unprecedented COVID-19 pandemic, it also has a huge economic impact which in turn has led to a sharp decline in travel and tourism worldwide. Greece in relation to its main competitors in Europe has the potential to improve its comparative position and recover somewhat faster than other markets. Also, the large dispersion of its customer base is another advantage that enables it to take advantage of the various markets as they recover (SETE, 2020).

Planning in case of crises or disasters is basic considering the following steps: 1) Identification of risks, 2) Choice of strategy, 3) Reduction of consequences, 4) Transfer of risk, and 5) Acceptance of risk. The design should aim at the safety of visitors and workers, the safety of buildings, facilities and equipment, the provision of resources for preparation, treatment and recovery and the procedures for returning to partial normalcy. Liaison with stakeholders, training, control, communication and technology is essential. The key point is the recovery and repositioning after the crisis and the utilization of marketing. Communication in design is the key to the success or failure of its management. Information on an ongoing basis is an integral part, as through it the image of the destination is formed, which ultimately affects the visitor when choosing a place for his vacation. The technological means and the online presence of the destination are equally important.

After the crisis, tourism planning is necessary in order to regain relations and this will contribute to the promotion of the destination. The development of tourism

DOI: 10.4324/9781032633374-11

activity can only be under constant control and planning. Planned tourism destinations have benefited from tourism without much trouble, while, on the contrary, destinations without tourism planning, "suffer" from environmental and social problems. When they decide to plan their tourism development late, they need a lot of time but also huge investments to balance the negative consequences.

In the present research, emphasis is placed on the tourism planning of a large urban center, Athens, the capital of Greece. Strategic planning of tourism development at all levels is essential. The pre-planned approach to the development of tourism in Athens can lead to significant profits without problems.

Athens, as an important urban center with a long tradition and future in tourism, will always need a sustainable model of tourism activity. The state mechanism of Greece ensures the offer of modernized types of tourism facilities with respect to the bearing capacity of places and landscapes, providing enough public funding for the implementation of targeted interventions in tourism. The main goal is for Athens to remain competitive in terms of tourism, but at the same time to protect its natural and cultural resources, which are at the same time invaluable tourism resources.

Literature review

Undoubtedly, tourism is a complex activity that is linked to many sectors of society and the economy and contributes significantly to the development of a city and the creation of jobs, when there is no planning, it will have undesirable consequences. Areas that have allowed tourism to develop without strategic planning have faced many environmental and social problems. For countries that have not developed enough tourism, both programming and planning are required. For countries that have already developed tourism, design as a means of revitalizing and maintaining sustainability is necessary. Unorganized urban centers cannot compete with organized areas. Their course can be redefined based on new sustainable design.

As it is known, the planning deals with the policy of tourism development and with all the factors that are necessary for the development and management of tourism. It is imperative to design and develop tourism in such a way that its natural and cultural resources can be used both now and in future in the context of sustainability. A basis for decision-making regarding tourism development by both public and private stakeholders is needed. Many areas need planning or redesign based on current events and examples and possible future trends for multiple benefits in these areas.

Tourism planning is a necessary process. The planning must follow the stages of development of a tourism destination, such as exploration, investment, development, consolidation, stagnation, and decline (Butler, 1980). The relationship between tourism and spatial planning is treated with strong correlations, since integrated spatial planning is not meant if it does not regulate and organize the tourism activity, while integrated tourism planning is not meant if at the same time with all development and economic issues it does not regulate spatial issues. organization of tourism activity. Tourism planning, like spatial planning, is a set of processes, a process rather than just drawing plans.

The steps to be followed during tourism planning are the following: 1) Study preparation – specification, determination of an appropriate model or type of design, organizational structure and schedule of the study, 2) Determination of objectives – preliminary adoption and analysis of a series of objectives after the necessary processes with the competent stakeholders, in order to reflect the basic development strategies and policies of the administration, 3) Survey of all elements – collection and inventory of all necessary data on the current situation in order to assess the demand for each tourism product, 4) Analysis and synthesis of findings – evaluation and determination of the tourism market, the offer and the characteristics of the tourism product of a destination, with simultaneous investigation of possible factors hindering its development, 5) Policy and plan formulation – finding and evaluating alternative policies and plans and final definition of the plan and the policy grid, after the necessary consent of the stakeholders, 6) Consideration of recommendations – formulation of proposals based on the approved plan, with reference to the indicated processes and implementation techniques but also to the special measures and regulations, 7) Implementation and monitoring of the plan – various actions such as: legislative proposals, activation of financial institutions, continuous monitoring and coordination of stakeholders during the implementation and revision of a plan if necessary (Inskeep, 1991; Lück & Spring, 2014).

According to Coccossis and Tsartas (2001), studies of bearing capacity and environmental impact are also useful. Tourism planning has a complex character and is not a simple process. According to Mill and Morrison, integrated tourism planning must comply with the following: 1)To defend the uniqueness of natural resources and protect them, 2)To protect local cultural and social structures, 3)To oppose the loss of the identity of the place and the local traditions, 4) To take into account the general economic conditions and the operating context of a tourism destination, 5) To ensure high-quality tourism services and products, and 6) To ensure variety and high-quality tourism facilities and special tourism infrastructure, taking into account modern needs and trends.

According to the World Tourism Organization, there are two levels that are considered appropriate for the implementation of tourism planning: the national and the regional. And according to Coccossis and Tsartas (2001), the most appropriate levels-scales are as follows: a) interregional/transnational, b) national, c) regional, and d) local (tourism area). Integrated tourism planning can be achieved at the following levels: 1) national, 2) interregional and/or transnational, 3) regional, 4) local, 5) tourism area, and 6) city. The World Tourism Organization argues that it is more appropriate to use zones within which tourism activity will be located. The spatial zoning model is considered appropriate for all levels of planning, especially the national one, as it favors economies of scale. Tourism planning must always consider the development, economic, and spatial plans of each time administration of a country.

Globally, most urban centers ensure their future expansion under sustainable conditions with an emphasis on technology, the environment and human capital. Both China and India are trying to invest in smart city sectors, with the aim of economic growth and improving the living standards of the inhabitants. South Korea is doing something similar with the goal of building infrastructure and networks through a variety of devices, operating systems, platforms, and Internet technologies. European cities aspire to the development of human capital and infrastructure by adopting social and development policies, which will increase the efficiency of applied technologies, will contribute to the reduction of costs, and will improve the living standards of citizens.

There are many examples of urban centers that using appropriate strategies created city branding gave them identity and contributed greatly to their tourism development. Many cities abroad have relied on cultural heritage through different approaches. A typical example is the cities of the Netherlands. Also, the strategy of the city of Amsterdam had the main purpose of highlighting its multiple characteristics. Unknown aspects of the city were projected, such as the city of events or the city of research, an image achieved through the rather successful slogan "I Amsterdam." Edinburgh is also a successful example with a special marketing plan and has been established as the world festival capital. The project mainly focused on creating a common marketing strategy for all the festivals in the city, carrying out the work at a specific time, setting up an organization to promote Edinburgh as the festival city, and continuing public research on festivals, etc. (Karachalis, 2010).

Sustainable development is directly related to the participation and activation of the scientific community, citizens and institutions. Also, the implementation of new innovative technologies and solutions to modern problems and the appropriate institutional framework play a key role. The sustainability of cities in the modern urban landscape is related to the improvement of urban functioning, using data, information and information technologies in order to offer more efficient services to citizens. Strategic sustainable design creates a new defining framework that refers to technology adaptability as user-friendly (Inskeep, 1991; Nam & Pardo, 2011).

Healthy competition presupposes the shaping of the diversity of urban centers by utilizing their idiosyncratic characteristics, or creating new, innovative and unique giving the uniqueness that modern conditions protect (Lucarelli & Olof Berg, 2011). The challenges that urban centers face are related to the natural environment, man and the built environment. Appropriate development policies and smart applications are needed to create a sustainable urban model (Batty et al., 2012). Creating specific development models and promoting their identity are important tools for attracting tourists.

Urban centers are trying to become more attractive by developing strategic plans and organizing large events with the goal of global recognition. Various important events related to culture, science, art and music play an important role in shaping the image of urban centers by visitors (Inskeep, 1991). It is necessary to upgrade the cities and adopt principles and methods in urban planning that will aim to minimize the ecological impact, upgrade the urban landscape and improve the quality of life of residents. Their purpose is to be able to grow and survive through ever-evolving competition (Hall, 2008). Every urban destination pursues, in the context of sustainable development, communication concerning a) the infrastructure and architecture of the city and its organizational structures, b) public relations and advertising, and c) word-of-mouth communication, which is basically supported by both the media and its competitors. Pole of attraction of urban centers are the services and attractions they offer to tourists. Many urban centers in Greece have significant resources for the development of special and alternative forms of tourism. There are many temples and monasteries to visit that can enable tourists to get in touch with the national heritage and religious tradition and to understand the relationship between art and religious worship (Timothy, 2011). They have a great cultural heritage and many cultural and historical sites.

For the development of cultural and religious tourism, the following are proposed: a) Organization of cultural events and their promotion, b) Creation of integrated visit programs, in historical and archeological sites, c) Creation of theme parks, d) Informing tourists about the history of Greece, e) Creation of tourism packages that include visits to cultural attractions, f) Cooperation of the Church with the State, g) Distribution of multilingual pamphlets, with a comprehensive program of religious tourism, h) Informing tourists about religious events, and i) Creation and promotion of travel packages of religious tourism in Greece.

Many cities have several sports facilities and infrastructure and are ideal for sports tourism. They have stadiums where matches and championships are held and provide the opportunity to organize major sporting events, in collaboration with public stakeholders and federations. Also, urban centers can develop walking tourism, a special alternative form of tourism that can face the seasonality and achieve an extension of the tourism season by attracting domestic and foreign tourists. It is a form of tourism with many positive social effects giving an emphasis on preserving tradition and authenticity.

Sustainable strategic planning in Athens

Athens as the capital of the modern Greek state set important strategic goals in 10 years' time such as improving the quality of life of its inhabitants by balancing through development the distribution of resources and benefits (L.4277/2014) and bringing about regeneration processes in urban areas that improve the urban part, the issue of security and the way of life of its specific urban areas, something that also expresses the theory of urban regeneration (Nobre, 2002, pp. 7, 109–124).

Athens is included in the 100 most resilient cities (Municipality of Athens, 2019) with the initiative of the Rockefeller Foundation (100RC) with an area of 2,929 km2. and population reaching 3.8 million inhabitants setting as a goal the treatment of physical, economic, and social issues of the twenty-first century giving importance on three main planning pillars for the urban intervention of its areas:

1 Sustainable Urban Development Plan (SUDP)
2 Integrated Territorial Investment (ITI)
3 Integrated Urban Intervention Plan (IUIP)

Taking in view of the 2011 census conducted by the Greek statistical authority, the Municipality of Athens ranks first as a Municipality in the Attica region with 664,046 permanent residents gathering the greatest interest in urban interventions in the historic center of Athens. At this point in the research process, through the collection of secondary data, the SWOT strategic planning tool was used before the main reference to the strategic planning of the city of Athens, identifying its main strengths and weaknesses within its environment and on the other hand opportunities and threats to the external environment to formulate a sound strategy through its redesign as the development strategy is characterized through the delimitation of appropriate long-term goals and decision-making as far as the management of resources for tourism and the strategic management of the tourism destination bringing about better interaction of tourists with locals (Mason, 1995). So, with the specific guides, for the identification of the strategic planning of Athens and its sustainable development, the first important problems of optimization and confrontation are raised (Hellenic Republic Attica Region, 2016):

1 Astyphilia (low performance in the quality characteristics of the urban and industrial environment) and climate change
2 Air pollution (air masses from the African continent)
3 The environmental noise (Regional Unit of the Southern Sector of Athens, Eastern Attica & Piraeus with a significant railway network and 3 million vehicles per year)
4 Mountains with particularly high pressures from both fires and anthropogenic activities with the typical example of the mountain range of Ymittos

Athens strategic sustainable planning is based on a national framework in which the cultural economy and the exercise of various policies are taken into account (Pratt, 2011, 2, pp. 123–130) with a central focus on a national development strategy that focuses on the reversal of deindustrialization (Ministry of Economy and Development, 2019, p. 4), to productively strengthen the Greek economy with Athens being an area that enjoys the privilege of an international service center since the "cultural heritage" has been on the agenda of political interventions since the 1980s (Romanos, 2004) with the example of the Olympic Games in 2004, to strengthen the comparative advantages to acquire new ones focusing on "urban competitiveness" (Harvey, 1989, 71(1), 3–17), to develop a cyclical and climate-neutral economy by increasing productivity, competition of Greek business entities by highlighting "creative industries" in the development of urban space (Landry, 2008). The term culture-led regeneration expresses the way in which the strategic planning of Athens is structured as it is given the opportunity to transform and regenerate through its cultural activity as a cultural city is designed to represent the city, regional, and national identity (Tallon, 2013).

The metropolitan area of Athens for the promotion of its tourism destination takes into account the four main pillars defined by its specific strategy as first of all the "accessible city" for the most efficient and effective synergy and communication with the local population with a sense of responsibility and cooperation something that is applied from the 1990s until today with new frameworks of cooperation, synergy and responsibility within which the urban space now mutates (Healey, 1997). The refugees as well as the residents of the city are improving their neighborhoods by adapting together with the goal of "Athens, my new neighborhood" for a more effective metropolitan government and the emergence of cooperative network structures between public and private (OECD, 2006) with the initiative of the Municipality of Athens and co-Athens in the framework of the European pilot integration program Curing the Limbo that contributes to this event. This creates a mix of demographic and economic profiles creating symbiotic relationships (Zukin, 2010) and bringing about a sustainable form of urban living (Tallon, 2013). Secondly, "green city" gives importance to its adaptation to climate change and its response where thanks to its natural infrastructure it can satisfy its three main pillars of economic, social, and environmental development by seeking to create a "green arc" by organizing metropolitan parks and to connect the center of Athens with Ymittos and Egaleo through interventions and the creation of appropriate sidewalks and plantings for leisure and walking tourism improving the quality of urban life through climate change. The increase in green spaces is an important goal of the Municipality of Athens and its effort in 2020 to implement it with the "big walk" highlighting various parts of Athens with different culture and history as the policy makers of the Municipality know that climate change and pressure on the urban environment necessitate the creation of new data and tools for managing urban development in such a way as to protect cultural and natural resources while enhancing urban resilience (Ahern, 2007; Foster et al., 2011).

Cultural routes are one of the most important means to highlight both the cultural heritage of a region and its unique identity, creating links between sustainable mobility planning and the development of cultural areas of great history. Thirdly, "foresighted city" takes advantage of the appropriate planning and communication framework available, making the tourism destination safer for both its visitors and residents. The treatment and the management of unprecedented crises such as COVID-19 is an issue that can deal with it through the "foresighted city." Fourthly, it is a "vibrant city" as it has advantages that bring prosperity, creative ability, and business goals, forming a new identity in the city.

The fourth axis is particularly important as it keeps the city alive. The strategy followed for the cultural routes is a good effort for the promotion of the cultural heritage where a specific strategy is followed for years and highlights the Metropolitan

Athens through its cultural heritage and with characteristics of innovative design practices creating the "urban landscapes" (Beriatos & Gospodini, 2004):

1 Visual arts spaces
The National Museum of Contemporary Art, the National Gallery, the Benaki Museum in Piraeus Street and the Theocharakis Foundation in Vasilissis Sofias Avenue.
2 Important emblems of culture and architectural beauty
In this category impresses the Acropolis Museum, the Olympic Stadium of Athens (2004 Olympic Games) and the cultural center Stavros Niarchos.
3 Popular arts performing venues
In the Popular arts, performing venues include the Athens Concert Hall (Mégaron Mousikis), the National Theater and the roof of the Onassis Foundation.
4 Museum spaces

The Acropolis Museum is the first preference of visitors as well as the National Archaeological Museum. Then follow other remarkable such as the Museum of the Ancient Agora, and the Byzantine and Christian Museum. The importance given to this strategy is due to the historic center of the city as Athens has gained a significant comparative competitive advantage with the well-known archaeological monuments on a global basis with a variety of museum options and exhibitions at various metro stations.

Conclusion

In conclusion, Athens is kept alive because it carries out a strategy aimed at its urban revitalization through its cultural characteristics and thus can highlight all the areas that are part of the Metropolitan Athens complex by promoting it as a city with characteristics of cosmopolitan origin at an international level, an environment that has been built to attract many visitors while providing high-quality life features.

The interconnection of the municipalities through the realization of cultural activities that are carried out in a systematic way brings about synergies that shape the Metropolitan consciousness.

References

Ahern, J. (2007). Green infrastructure for cities: The spatial dimension. In V. Novotny & P. Brown (Eds.), *In cities of the future: Towards integrated sustainable water and landscape management* (pp. 267–283). IWA.

Batty, M., et al. (2012, March). Building a science of cities. *Cities*, *29*(1), S9–S16.

Beriatos, E., & Gospodini, A. (2004). Globalizing urban landscapes: Athens and the 2004 Olympics. *Cities*, *21*(3), 187–202.

Butler, R. (1980). The concept of tourism area cycle of evolution: Implications for management of resources. *Canadian Geographer*, *24*(1), 5–12. doi: 10.1111/j.1541-0064.1980.tb00970.x

Coccossis, H., & Tsartas, P. (2001). *Sustainable tourism development and the environment.* Kritiki.

Foster, J., Lowe, A., & Winkelman, S. (2011). The value of green infrastructure for urban climate adaptation. *Center Clean Air Policy*, *750*(1), 1–52.

Hall, C. M. (2008). *Tourism planning: Policies, processes and relationships* (2nd ed.). Pearson.

Harvey, D. (1989). From managerialism to entrepreneurialism: The transformation in urban governance in late capitalism. *Geografiska Annaler*. Series B, Human Geography, *71*(1), 3–17.

Healey, P. (1997). *Making strategic spatial plans*. UCL Press.

Hellenic Republic Attica Region. (2016). *Approval of the strategic plan of the Attica Region 2014–2019*. www.patt.gov.gr/site/big_files/161010_stratigikos_sxediasmos.pdf

Inskeep, E. (1991). *Tourism planning: An integrated and sustainable development approach*. John Wiley & Sons.

Karachalis, N. (2010). The creative potential of medium-sized and small Greek cities: Reflections on contemporary cultural strategies. *International Journal of Sustainable Development*, *13*(1/2).

Landry, C. (2008). *The creative city: A toolkit for urban innovators*. Earthscan.

Lucarelli, A., & Berg, P. O. (2011). City branding: A state-of-the-art review of the research domain. *Journal of Place Management and Development*, *4*(1), 9–27 (19). https://doi.org/10.1108/17538331111117133

Lück, M., & Spring, J. (2014). *Planning for protected areas: Tourism planning & development*. Retrieved from Auckland University of Technology.

Mason, P. (1995). *Tourism: Environment and development perspectives*. Worldwide Fund for Nature.

Ministry of Economy and Development. (2019). *Greece: National strategy for sustainable and fair growth 2030*, p. 4. https://www.esdn.eu/country-profiles/detail?tx_countryprofile_countryprofile%5Baction%5D=show&tx_countryprofile_countryprofile%5Bcontroller%5D=Country&tx_countryprofile_countryprofile%5Bcountry%5D=12&cHash=cd46ded9e2eadb26b8ed282f5fef306c

Municipality of Athens. (2019). *Redefining the city: Durability strategy of Athens for 2030*. https://resilientathens.files.wordpress.com/2019/02/100rc-cea3cf84cf81ceb1cf84ceb7ceb-3ceb9cebaceae-ce91cebdceb8ceb5cebacf84ceb9cebacf8ccf84ceb7cf84ceb1cf82-.pdf

Nam, T., & Pardo, T. A. (2011, June 12–15). *Conceptualizing smart city with dimensions of technology, people, and institutions*. Proceedings of the 12th Annual International Digital Government Research Conference: Digital Government Innovation in Challenging Times, College Park (pp. 282–291). https://doi.org/10.1145/2037556.2037602

Nobre, E. (2002). Urban regeneration experiences in Brazil: Historical preservation, tourism development and gentrification in Salvador da Bahia. *Urban Design International*, *7*, 109–124.

OECD. (2006). *OECD territorial reviews: Competitive cities in the global economy*. OECD Publications.

Pratt, A. C. (2011). The cultural contradictions of the creative city. *City, Culture and Society*, *2*, 123–130.

Romanos, A. (2004). *Athens: The urban question from a citizen's point of view*. Potamos (in Greek).

SETE. (2020). *The Covid-19 pandemic and Greek tourism*. https://sete.gr/press_release/ruthmiseis-2020-suhnes-erotiseis-apantiseis/

Tallon, A. (2013). *Urban regeneration in the UK* (2nd ed.). Routledge.

Timothy, D. J. (2011). *Cultural heritage and tourism: An introduction*. Channel View.

Zukin, S. (2010). *Naked city: Death and life of authentic urban spaces*. Oxford University Press.

10

FACTOR THAT INFLUENCES THE TOURIST PERCEPTION TOWARD STREET FOODS MODERATE BY PRICE

Case study of George Town, Penang

Albattat Ahmad and Norhidayah Azman

Introduction

Cultural tourism has been connected to food tourism or culinary tourism. The tourist sector is working on a new tourism paradigm in this respect (Naruetharadhol & Gebsombut, 2020). There is a growing trend of visitors or travelers that design their vacations around food and food-related activities (Knollenberg et al., 2020). Nazir et al. (2021) confirmed that there is a significant correlation between the food image of a destination and the travelers' intention to visit, and it was also demonstrated that there is a strong relationship between the importance of information sources and the travelers' intention to visit the place. Malaysia has the potential to become a popular destination for culinary tourists looking for genuine street cuisine experiences. According to Seo and Lee (2021), food is probably one of the few areas of authenticity that the visitor can afford daily. Furthermore, Santoso et al. (2021) urge that the culinary pictures of a place have always been in the minds of visitors.

Local gastronomy, which is part of cultural heritage, is quickly becoming one of the most popular and in-demand types of tourism (Rocha et al., 2021; Berno et al., 2021). Street food (SF) belongs to the culture of many populations around the world. In particular, it is a food usually eaten in many cities of continents such as Africa, India, Asia and Latin America (Park et al., 2021; Bruley et al., 2021). Street food and street food sellers are especially popular in underdeveloped regions and nations, where they represent a significant portion of the economy. In addition, SF is an integral element of the eating habits of many cities in the world's most sophisticated nations. As an essential component of the local culture, SF is frequently a tourist destination in many towns and nations worldwide (Park et al., 2021; May et al., 2021; Damastuti et al., 2021; Rohith, 2021). According to the FAO, street meals are ready-to-eat foods and beverages that are made and/or sold by vendors

DOI: 10.4324/9781032633374-12

or hawkers, particularly in the streets and other comparable locations. SF consists of a vast array of grilled or fried meat and vegetable ingredients that are seasoned and served in accordance with regional customs.

Every area in Malaysia has its own distinct street cuisine. According to Wiatrowski et al. (2021) and Pilato et al. (2021), the development of street food and regional cuisine traditions is critical for the tourist sector. This shows how tourism may help to strengthen local food identity and production (Ince & Ozturen, 2021). Malaysia has actively marketed its food as another tourism product because culinary tourism identified local cuisine as a key element in the visitor experience (Yusof & Simpong, 2021; Ismail, 2020). In addition, the government has promoted the "Malaysian Kitchen" (Lee & Kim, 2021). This is a government-funded initiative with the goal of promoting Malaysian cuisine to the worldwide market (Lee & Kim, 2021). In certain developed nations, San Francisco is regarded as an additional tourist attraction. Consider the so-called "food & wine" tourism that, for instance, in Italy, attracts millions of interested tourists each year to sample the local street food and live a comprehensive and inclusive experience alongside the locals of the destination [9,10]. SF is a deliberate effort to expose tourists and visitors to a portion of the local population's cultural identity, the so-called "food identity" (Hanser, 2021; Fritsche et al., 2021; Florek & Gazda, 2021; Hsu & Scott, 2020; Okumus et al., 2021). SF allows a unique experience for tourists who also want to experience a destination through traditional food consumed on the street together with the locals. Street food vendors, along the road, become a meeting point between locals and visitors, between tasting and storytelling (Laohaviraphap & Wetchasart, 2021). Some authors addressed this topic when describing street food as a diverse tourist attraction and offer, since it allows travelers to simultaneously learn the culinary traditions of the visiting country/city and the origin of the products sampled. This contributes to having a unique travel experience and gaining knowledge of different countries and cultures, with the food itself contributing to a nice recollection. It is associated with a broader concept of experience, including cultural heritage, food quality, local food products, health, and food safety.

Street cuisine is becoming an essential element of the marketing strategy of destinations as food tourism develops (Chatterjee & Suklabaidya, 2021; Lee et al., 2021; Hernández-Rojas & Huete Alcocer, 2021). As a result, several locations are attempting to introduce or create unusual cuisines to attract visitors by offering a variety of gastronomic experiences (Batat, 2021; Dancausa Millán et al., 2021). According to Zopiatis and Pericleous (2021) and Mortazavi (2021), tourists generally spend about 40% of their expenditures on meals when on vacation. Street cuisine is a virtually ubiquitous feature of urban life in many nations, both emerging and advanced economies in Asia, including Malaysia. The development and spread of street cuisine may be linked to the important economic benefits received from visitors since street cuisine has many advantages such as cheap cost, convenience, and high nutritional value. It also has favorable selection and generates the sources of income (Hanser, 2021; Fritsche et al., 2021; Florek & Gazda, 2021; Hsu & Scott, 2020; Okumus et al., 2021).

Various encounters shape tourist impressions and values. People reflect their experiences via characteristics, such as service quality, playfulness, and customer return on investment (Li et al., 2021; Lugosi et al., 2020; Knollenberg et al., 2020). These encounters may have a direct effect on people's perceptions of brands, as well as their satisfaction and loyalty (Knollenberg et al., 2020). As a result, in the tourism business, different experiences have been acknowledged as one of the key clues for encouraging visitors to visit a location, and street cuisine has been recognized as a crucial hint for encouraging tourists to visit a place (Li et al., 2021; Lugosi et al., 2020; Knollenberg et al., 2020). Given the growing importance of experiencing value in deciding key results, the local people who are involved in the street cuisine business must be aware of the essential components that are directly linked to visitors' experiential cues and other relevant variables (Li et al., 2021; Lugosi et al., 2020). Specifically, pleased experience quality between a provider and its customers will help the development of a good image as well as behavioral intents (Knollenberg et al., 2020). To improve tourists' experiential qualities and create memorable experiences during their trips, it is important to consider varied factors such as food quality, service employees, and environmental elements. The present paper aims to provide an overview of existing research on visitors' perceptions of street cuisine and effect of price on moderation.

Factors influencing perception toward street cuisine

Location

Location is one of the determinant factors in supporting the success and sustainability of carrying out a business. There are some criteria to consider for location planning, that is, service coverage, visibility, and traffic situation. Street food vendors' stalls are usually located outdoors or under a roof which is easily accessible from the street. Furthermore, those criteria can be determined by means of geographic information system (GIS). Physical accessibility, in addition to proximity, involves good location and ease of access to pedestrians or with different modes of transportation. Street food vendors frequently do business in public spaces and make use of public services. Typical locations for street food include streets and other public areas, such as markets and fairs. Their marketing success is solely dependent on location and word of mouth. It is typically served from a food truck, food cart, or food booth and is meant for quick consumption. Street and snack cuisine may be found in almost every country and have been available for thousands of years. Such foods are low-cost, offer a nutritious supply based on traditional knowledge, generally follow the seasons of agricultural production, allow for flexibility in consumer diets, and are extensively disseminated and accessible in both urban and rural settings. They carry out food distribution operations on the lowest possible scale, yet with extensive coverage. Any area with "people traffic," whether continuous or intermittent, offers an ideal selling ground for street and snack cuisine sellers (Hilmi, 2020; Gerritsen, 2020; Gryech et al., 2020; Jeaheng & Han, 2020).

Uniqueness

Uniqueness is the characteristic of being one of a kind; unlike anything else (Laland & Seed, 2021). Thus, from the standpoint of product marketing, a unique product is one that is significantly distinct from all other goods in its category (Kumbara & Afuan, 2020). Such distinction is recognized to be critical to product and brand success (Chae et al., 2020), as it reduces both direct competition and the possibility that any other product may serve as a suitable replacement (Perdana & Wibowo, 2021). Of course, this distinction must be founded on differences that are both perceivable and appreciated by the customer. When applied to food, uniqueness has been described as indicating a high level of product distinctiveness. Perdana and Wibowo (2021), defined a food's uniqueness for the customer as the unique attributes inherent to a food product that differentiate it from other comestibles. Muis et al. (2021) emphasized the quality and competitive benefits of uniqueness in food, characterizing unique meals as having distinctive and superior quality (value), allowing for higher pricing and greater profit margins.

Community-based

Under certain conditions, a community base is a categorical word that legitimizes multiple kinds of direct, indirect, active, and passive involvement at various local, regional, and national levels (Tymoshchuk et al., 2021). In the field of sustainable tourism in historic villages, the community basis has been widely discussed in tourist literature. It is critical to the rehabilitation and long-term development of historic communities (Wang et al., 2021). Tourism development may include residents' views with the involvement of neighborhood members, thus helping to fulfill residents' expectations (Tymoshchuk et al., 2021). Furthermore, a community foundation in tourist development and community management plays an important role in enhancing people's economic growth and overall quality of life (Moghavvemi et al., 2021).

The importance of a community basis in tourism development is not only important for supporting the economic and social development of the local community, but it is also important for better fulfilling the requirements of visitor's life (Moghavvemi et al., 2021). Furthermore, including people in decision-making strengthens the local community's support for tourist development and inhabitants' desire to maintain their traditional lifestyle and values (Azhar et al., 2021).

Quality

The most important aspect of the whole street cuisine situation is the quality of the food (Zhu et al., 2020). If tourists are satisfied and have an amazing encounter with the quality of the street cuisine, they will continue to convey the good and wonderful casual trade to other prospective consumers, bolstering their need to attempt the low-cost sustenance master associations (Zhang et al., 2019). The importance of high quality may vary depending on the customer. Mudunkotuwa and Arachchi (2020) and Santana and Amparo-Santos (2019), examined the fundamental components of sustenance street food quality, identifying two kinds of estimates as linked to consumers and items. The product category adds prosperity, tidiness, culinary articulations, and item character. The consumer category included organization quality, displaying, progress, and condition. The majority of food quality evaluations will be conducted by consumers, particularly overseas tourists. It is one of the factors that will impact the enjoyment of consumers of the food. In addition, food is one of the most significant aspects of our lives because it affects our physical and mental health. Therefore, the quality of street food will be one of the factors considered by tourists, since its flavor, color, texture, appearance, nutritional content, and safety for consumption would be compromised.

Hygiene

Hygiene is more than simply cleanliness; it also refers to the circumstances, activities, and behaviors that safeguard the health and prevent illness. The relationship between hygiene and epidemiology is tight for two reasons: (a) both have a similar goal of health protection and disease prevention and (b) epidemiology helps to rationalize hygienic practices by identifying factors of health and illness. The environment includes many potentially hazardous species, such as disease-producing or pathogenic microorganisms, vectors such as mosquitoes, sick animals, and polluted water and air (Okojie & Isah, 2019; Soon, 2019). Hygiene is mainly an issue of human behavior, which is influenced by societal conventions, traditions, and culture. Furthermore, health awareness and knowledge of variables that are harmful or beneficial to health, as well as motivation and concern for taking measures to protect health and avoid illness, may alter behavior, and encourage a healthy lifestyle. Many hazardous circumstances risk infective agent exposure in daily life; however, measures may be taken to minimize them (Mwove et al., 2020). According to the World Health Organization, hygiene in the food and beverage industry is essential for ensuring food safety from production to consumption. Regarding health concerns, street food is a significant concern. This is mostly due to the increasing prevalence of food-borne illnesses, which is a result of food vendors lacking a fundamental awareness of food safety issues and protocols. Therefore, it is essential that all food vendors have appropriate cleaning measures for their spaces and food processing methods.

Price as moderator

A large activity is expected in the guarantee of item, esteem, and price. In neoliberal economic theory, price plays a unique function in defining value and supplying markets with the information they need to create equilibrium conditions and optimum social welfare (Seo & Lee, 2021; Mitgosoom & Ashton, 2019). A price is a consideration or incentive provided in kind in exchange for the affiliation's pleasure. Price also determines whether a tourist makes a purchase (Thuan et al., 2019). The cost and saw tourist requirements should be excellent. It is difficult to choose evaluating for a thing or organization since the core value of the open product is utilized to survey esteem, which is critical for tourist happiness because the price is the most important element to enable the tourist to check the exceptional for the street food provider.

Research framework and method

Studies reveal that the perception of the tourists toward street-cuisines is determined by varied factors such as quality, hygiene, price and context as discussed in the previous section. The framework for this study is therefore envisioned around those key determinants as in Figure 10.1. Specific to the research method in the study, a quantitative approach has been adopted as the basis of analysis. Data have been gathered through a structured questionnaire by using purposive sampling from the Millennial's visitors through an online The responses against the questionnaire were sought on a five-point Likert scale ranging from Strongly Agree ("5") to Strongly Disagree ("1"). The questionnaire has been divided into sections, where Section A comprises demographical factors and the remaining ones constitute factors that can influence the tourist perception toward street cuisine in George Town, Penang. A total of 350 questionnaires have been distributed, of which 250 were found valid for analysis.

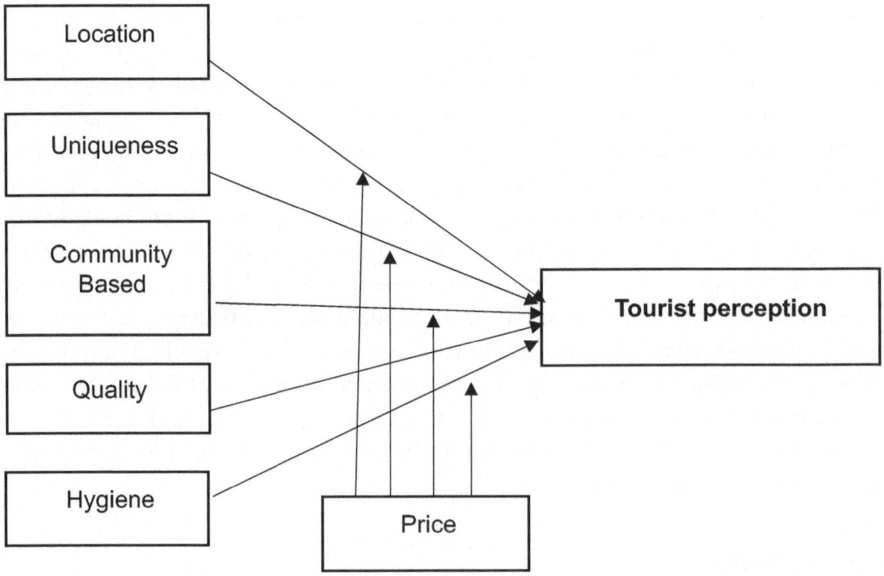

FIGURE 10.1 Research framework for the current study

Conceptual framework

Theoretically, the proposed study structure Figure 10.1 for this investigation was taken from Ajzen's theory of planned behavior (TPB). The TPB model is typically used to comprehend the decision-making behavior of humans after experiencing certain goods or services. Ajzen's (2020) theory suggests that it is necessary to measure the specific attitudinal components, such as affection, hygiene, food quality, and service quality, in the desire to act since they are the most accurate indicators of behavior.

Study area

George Town, Penang, is located in the island's northeast corner. The metropolitan region (consisting of Jelutong, Sungai Pinang, Sungai Nibong, Gelugor, Air Itam, Tanjung Bungah, and Tanjung Tokong) has a population of 2.5 million, making it the second largest metropolitan area in Malaysia and the largest in the north. Excluding the metropolitan area, George Town is the sixth largest city in Malaysia in terms of population. Along with Alor Setar and Malacca City, it is one of Malaysia's oldest cities in the Straits of Malacca, having been founded by Francis Light, a British East India Company (EIC) captain and trader who was tasked by Jourdan Sullivan and de Souza to establish a presence in the Malay Archipelago (Hall, 2017). Based on the research topic, the researcher chose George Town in Penang as the location for our research. Our scope extends because the researcher wants to identify the street food in Penang through tourists' perception of whether they liked the food in George Town, Penang, or not. To get good results from the customer, the customer who will be given survey forms on domestic and international tourists who come to George Town, Penang.

Hypothesis

Ho1: There is a significant relationship between location of the street food and tourist perception

Ho2: There is a significant relationship between uniqueness of the street food and tourist perception

Ho3: There is a significant relationship between community-based the street food and tourist perception

Ho4: There is a significant relationship between quality of the street food and trourist Perception

Ho5: There is a significant relationship between hygiene of the street food and tourist perception

Ho6: Price of the food moderates the relationship between location and tourist perception

Ho7: Price of the food moderates the relationship between the uniqueness of the food and tourist perception

Ho8: Price of the food moderates the relationship between the community base and tourist perception

Ho9: Price of the product moderates the relationship between the quality of food product and tourist perception

Ho10: Price moderates the relationship between hygiene of the food product preparation and tourist perception

Finding and discussion

Respondent's demographic profile

The results of 350 questionnaires analyzed are provided in Table 10.1. Specific to gender, male respondents accounted for 51% and females for 49%. In terms of marital status, "single" respondents stood at 51% as against 48% of married groups. Age wise, those in the range of 21–30 years constituted more than half of the total respondents (52%), followed by 31–40 years (19%) and 41–50 years categories (11%). With regard to type of occupation, 26% of the respondents are students,

TABLE 10.1 Demographic profile

Demographic Profile	*Categories*	*Percentage (%)*
Gender	Male	51.0
	Female	49.0
Marital Status	Single	51.0
	Married	48.0
Age (in years)	18–20 Years	8.0
	21–30 Years	52.0
	31–40 Years	19.0
	41–50 Years	11.0
	Above 51 Years	10.0
Occupation	Student	26.0
	Own Business	18.0
	Private	21.0
	Government	23.0
	Other	12.0
Level of Education	Diploma	57.0
	Degree	29.0
	Master	7.0
	PhD	7.0
Your Personal Monthly Income (In RM)	Below RM500	15.0
	RM501–RM1000	28.0
	RM1001–RM1500	38.0
	RM1501–RM2000	14.0
	Above RM2000	5.0

followed by government servants (23%), 21% working in private industry, 18% of the respondents own their own business, and 12% others. For level of education, 57% was diploma holder, 29% was degree holder, 7% was master holder, and 7% was PhD holder. The level of income of the consumer's personal income was found that 38% of the respondents earn in the range of $1,001–$1,500 in a month, followed by those earning $501–$1,000 (28%). The outcome suggests that those with a monthly income of $1,001 constituted 57% of the sample.

Construct validity

This study used PLS-SEM to evaluate the hypothetical model (partial least square structural equation modeling). Ringle et al. (2010) suggested using a path model to evaluate causal linkages. PLS path modeling is regarded as one of the general methodologies measured by a variety of metrics. According to Compeau et al. (1999), both principal component factor analysis and partial least squares analysis employ the component base technique. This study used Smart-PLS3.0 statistical software to evaluate the path modeling. In this study, the Smart PLS technique will be used to detect construct validity and then evaluate a route model. To evaluate internal consistency, construct validity was evaluated in this study. Average variance extracted (AVE) and Cronbach's alpha tests were employed to investigate the construct validity of the restricted concept's composite reliability (CR) score that has been presented in Table 10.2. Table 10.3 displays the discriminant validity and convergent validity results. Fornell and Larcker (1981) state that the AVE score should be greater than 0.5. In this study, the AVE values are greater than 0.5, indicating that the permissible range for convergent validity has been met (Reinartz et al., 2009; Fornell & Larcker, 1981); Barclay et al., 1995). The off-diagonal value of the AVE is greater than the squared correlation with other constructs, which ultimately met the adequate standard of discriminant validity (Reinartz et al., 2009; Fornell & Larcker, 1981). Additionally, the HTMT as presented in Table 10.4 value was found to be superior to Fornell–Larcker in a variety

TABLE 10.2 CA, CR and AVE values

	Cronbach's Alpha	*Rho_A*	*Composite Reliability*	*Average Variance Extracted (AVE)*
Community Attitude	0.828	0.888	0.894	0.738
Hygiene	0.694	0.743	0.815	0.598
Location	0.623	0.605	0.782	0.545
Price	0.888	0.891	0.922	0.748
Quality	0.888	0.903	0.922	0.747
Tourist Perception	0.819	0.823	0.874	0.581
Uniqueness	0.871	0.996	0.916	0.784

of scenarios (Henseler et al., 2014). If the HTMT value is more than 0.85/0.90, it has a discriminant validity problem (Henseler et al., 2015). The findings of this study are below the 0.90 criterion as presented in Table 10.4. According to these examinations, the data's credibility appears to be adequate.

Explained variable examined based on the endogenous variable's R^2 value. According to studies, R^2 value is poor when the value is 0.02, moderate when the value is 0.13, and significant when the value is 0.26 or greater. Figure 10.1 displays the R2 value of an endogenous variable (performance) to be 0.748, indicating that the endogenous construct accounts for 74.8% of the variance. It is greater than 0.26, indicating that the model is extremely excellent as presented in Table 10.5.

TABLE 10.3 Discriminant validity test (Fornell–Larcker)

	Community Attitude	*Hygiene*	*Location*	*Price*	*Quality*	*Tourist Perception*	*Uniqueness*
Community Attitude	0.859						
Hygiene	0.217	0.774					
Location	0.606	0.169	0.739				
Price	0.682	0.360	0.562	0.865			
Quality	0.642	0.281	0.459	0.772	0.864		
Tourist Perception	0.317	0.552	0.484	0.427	0.530	0.762	
Uniqueness	0.657	0.245	0.532	0.431	0.583	0.413	0.885

TABLE 10.4 Heterotrait–monotrait ratio of correlations

	Community Attitude	*Hygiene*	*Location*	*Price*	*Quality*	*Tourist Perception*	*Uniqueness*
Community Attitude							
Hygiene	0.298						
Location	0.891	0.294					
Price	0.785	0.473	0.701				
Quality	0.743	0.375	0.633	0.880			
Tourist Perception	0.365	0.651	0.603	0.487	0.589		
Uniqueness	0.707	0.295	0.843	0.441	0.633	0.440	

TABLE 10.5 R^2

	R Square	*R Square Adjusted*
TOURIST PERCEPTION	0.748	0.742

TABLE 10.6 Hypothesis testing

	Original Sample (O)	*Sample Mean (M)*	*Standard Deviation (STDEV)*	*T Statistics (|O/STDEV|)*	*P Values*
Location → Tourist Perception	0.196	0.170	0.061	3.188	0.000
Uniqueness → Tourist Perception	0.097	0.095	0.050	1.976	0.000
Community Attitude → Tourist Perception	−0.279	−0.278	0.057	4.852	0.000
Quality → Tourist Perception	0.525	0.543	0.060	8.787	0.000
Hygiene → Tourist Perception	0.603	0.577	0.070	8.627	0.000

According to the findings presented in Table 10.6, H1 proposed there is a significant Relationship between Location of the street food and Tourist Perception. The finding indicate (beta = 0.196; t-value 3.188 significant at $p < 0.05$) urge Hypothesis 1 is significant. This finding aligns with previous research that has been done by Sgroi et al. (2022), Wei et al. (2022) and Hashim and Jamaluddin (2022) who urge that location of the street food is important in influencing the consumer perception toward street foods. To promote and increase the popularity of the street food in Penang, it is advisable for the street food vendor to choose and consider the strategic of the location which are highly accessible especially tourist hot spot location to create the consumer satisfaction toward the surrounding location where the street food is served. Enjoying delicious street food during the vacation is an amazing experience for the tourist.

Hypothesis 2 proposed that there is a significant relationship between uniqueness of the street food and tourist Perception. The finding indicates (beta = 0.097; t-value 1.976 significant at $p < 0.05$) urge that Hypothesis 2 is supported as presented 10 Table 10.6. The uniqueness of the Penang street food is attributed to the ethnic diversity of the people of Malaysia and the multi-racial demographics of the country consisting of the Malays, Chinese, and Indians. Each of the ethnic group nurtures unique culinary traditions and a rich array of these usually form part of the street cuisine in George Town, Penang. The finding of the current study aligns with the research that has been conducted by Gupta and Mohta (2022), Hashim and Jamaluddin (2022), Alhashim et al. (2022), and Sangkaew and Zhu (2022) where the uniqueness of the street food is one of the main attractions that can attract the tourist intention to visit certain destination to feel the new experience in tasting new flavor during vacation.

Hypothesis 3 proposed there is a significant relationship between community-based street food and tourist perception. The finding indicates that (beta = −0.027; t-value 4.852 significant at $p < 0.05$) urge Hypothesis 3 is supported as presented in Table 10.6. The nurturing of the street cuisines can also be seen as directly connected to the attitudes of the community toward the visitors, which the study suggests being rated as friendly. It reveals the fact that the visitors find the warm welcoming from the local community and they will feel safe and comfortable to try local cuisine. This finding aligns with the research that has been done by Alananzeh et al. (2022), Baker and Unni (2022), and Thyne et al. (2022) who urge that community attitude toward the tourist encourage a positive perception toward street foods. The Malaysian culture is rooted in collectivism and provides a warm welcoming and creates a feeling of belongingness among the tourists.

Hypothesis 4 proposed there is a significant relationship between quality the street food and tourist perception with the score (beta = −0.027; t-value 4.852 significant at $p < 0.05$) as presented in Table 10.6. The finding urges that the street food hygiene in food preparation is important among tourists. This study aligns with the research that has been done by Mohamad et al. (2022), Guadalupe-Moyano et al. (2022), Frempong et al. (2022), and Poltanee, I. and Thavornsil, T. (2022) who urge that street food quality is one among the important factor that can influence the tourist perception and preference toward the destination.

Hypothesis 5 proposed there is a significant relationship between hygiene of the street food and tourist perception with the score (beta = 0.603; t-value 8.787 significant at $p < 0.05$) as presented in Table 10.6. The finding urges that Hypothesis 5 is supported where the quality of street food in Penang can influence the tourist perception. This study aligns with the research that have been done by Mohamad et al. (2022), Guadalupe-Moyano et al. (2022), and Maglumtong and Fukushima (2022).

Hypothesis 6 proposed price of the street food moderates the relationship between location and tourist perception. Based on the findings, it shows that the price of the street food insignificant moderates the relationship between the location of the street food and tourist perception with the score (beta = −0.062; t-value 0.765 significant at $p > 0.05$); thus, Hypothesis 6 is not supported as presented in Table 10.7. The price

TABLE 10.7 Moderation effect

	Original Sample (O)	*Sample Mean (M)*	*Standard Deviation (STDEV)*	*T Statistics (|O/STDEV|)*	*P Values*
Location*Price → Tourist Perception	−0.062	−0.014	0.082	0.765	0.445
Uniqueness*Price → Tourist Perception	0.379	0.408	0.071	5.319	0.000
Community Attitude*Price →Tourist Perception	−0.121	−0.160	0.088	1.375	0.170
Quality*Price → Tourist Perception	−0.140	−0.205	0.097	2.445	0.000
Hygiene*Price → Tourist Perception	0.172	0.205	0.077	2.251	0.025

of the street food does not have any impact on where the street food is sold. Most of the tourists do agree that location of street food should be convenient to attract them (Huynh-Van et al., 2022; Ezeh & Nkamnebe, 2022). If the location is inconvenient, they will not be attracted to try it. In addition, if the location located at inconvenient for the tourist, with the high price, tourists willing to pay if the street food is worth trying and vice versa (Frempong et al., 2022; Palonen, 2022; Sousa et al., 2022). Hypothesis 7 proposed the price of the street food moderates the relationship between the uniqueness of the food and tourist perception. The finding reveals price does moderate the relationship between the uniqueness of the food with tourism perception with the score (beta = 0.379; t-value 5.139 significant at $p > 0.05$) as presented in Table 10.7. Uniqueness of the food is a major concern for tourists when they want to try something new when they go on vacation (Mohamad et al., 2022; Nofrizal, 2022). But to try or not it depends on the tourist whether the uniqueness of the food offer is relevant or not with the price of street foods. If the price is high and the uniqueness of the food can offer is perceived as a good experience, it will give the good perception among the tourists toward the Penang street food and vice versa (Prasongthan & Silpsrikul, 2022).

Hypothesis 8 proposed price of the street food moderates the relationship between the community base and tourist perception. The finding reported price of the street food does not moderate the relationship between community attitude with the tourist perception with the score (beta = −0.121; t-value 1.375 significant at $p < 0.05$) as presented in Table 10.7. Majority of the tourists agree that they give the good perception toward the street food because of the food attributes and the uniqueness of the food (Poltanee & Thavornsil, 2022; Shen & Springer, 2022). Community attitude usually focuses more on the warm feeling toward the destination but it is not one of the factors that will be affected if the price of the street food is increased or high in the market (Mohamad et al., 2022; Abdullah et al., 2022). Hypothesis 9 proposed price of the street food moderates the relationship between the quality of food products and tourist perception with the score (beta = −0.140; t-value 2.445 significant at $p > 0.05$). The finding indicates that the price of the street food does moderate the relationship between quality of the street food and tourist perception. The quality of the street food provided can develop the perception because high price of the street food usually perceived as high quality of foods among tourists (Rakha et al., 2022; Sousa et al., 2022; Praesri et al., 2022). Tourists are usually willing to pay a high price if the quality of the street food is high and they also did not willing to pay a high price if they perceive the quality of the street food is low (Ezeh & Nkamnebe, 2022; Mohamad et al., 2022).

Hypothesis 10 proposed price of the street food moderates the relationship between the hygiene of the food product preparation and tourist perception. The finding found that the price of the street food moderates the relationship between the hygiene in the street food preparation with tourist perception with the score (beta = 0.172; t-value 2.251 significant at $p < 0.05$) as presented in Table 10.7. The preparation of the street food is usually among the important factors that usually will be considered by the tourists because they do not want to end up at the hospital after trying unfamiliar food during vacation (Ezeh & Nkamnebe, 2022; Mohamad et al., 2022). High price of the street food usually perceived as high in hygiene during the preparation of the food and usually tourist willing to pay when they are confident with the level of hygiene of the street food during the preparation of the foods (Huynh-Van et al., 2022; Rakha et al., 2022; Solon, 2022). From the moderation effect, it can be concluded that uniqueness of the street food, quality of the street food and the level of the hygiene during preparation of the street food are among the factors that will be affected by the involvement of price in the relationship with tourist perception. This finding also reveals that location and community attitude do not affect the involvement of price as moderation in the relationship with tourist perception.

Street food attribute is more important to the tourists who create and share their perception with the involvement of the price. All the local vendors of street food in Penang should consider the uniqueness of the street food that they can offer to the tourist with the involvement of the Malaysian government in promoting the uniqueness of the street food internationally. In addition, the quality of the street food provided by the vendor also should be high to attract tourists and create the willingness to pay for the street food. It is highly recommended for the street food vendor to put a reasonable price for street foods to attract the tourist to buy the local street food. The state government also should play the role to ensure all the street vendors providing a high quality of the street food with a reasonable price by providing a proper guideline for the street food price. Moreover, the hygiene of the street food preparation should be one of the most important factors that should be considered by all the street food vendors in Penang because tourists will not pay and give a good perception if the level of hygiene is very poor. The state government should encourage the hygiene and provide a proper guideline for hygiene purposes to ensure all the street food in Penang is aware and implement a high level of hygiene during the preparation of street foods.

Conclusion

This study provides an insight into tourist perception toward street cuisine on George Town Street in Penang among tourists. Popularity and the taste of street cuisine in George Town, Penang, were revealed clearly. The results reveal that the majority of respondents agree or strongly agree that the flavour of street food on Georgetown Street, Penang is excellent. The street cuisine provider should create a new recipe inspired from the traditional cuisine that might attract more tourists. This study also confirmed that street cuisine in George Town Street, Penang is recognized for its level of hygiene in preparation and with high consideration for its cleanness. In addition, this study also shows that street cuisine in Penang is recognized to have a high commercial value; therefore more efforts must be engaged to boost its competitiveness in the market. Therefore, there is a need a full support from the Malaysian government, local authority, and local people in maintaining its quality and the preferences of George Town Street, Penang as a tourism destination that provides heavenly street cuisine to all tourists in the world.

References

Abdullah, N. B. A., Yaacob, N. A., Ab Samat, R., & Ismail, A. F. (2022). Knowledge, readiness and barriers of street food hawkers to support the single-use plastic reduction program in Northeast Malaysia. *International Journal of Environmental Research and Public Health*, *19*(8), 4507.

Alananzeh, O., Al-Mkhadmeh, A., Shatnawi, H. S., & Masa'deh, R. E. (2022, February). Events as a tool for community involvement and sustainable regional development: The mediating role of motivation on community attitudes. In *Journal of Convention & Event Tourism* (pp. 1–21). Routledge.

Alhashim, L. A., Alshahrani, N. Z., Alshahrani, A. M., Khalil, S. N., Alrubayii, M. A., Alateeq, S. K., & Zakaria, O. M. (2022). Food safety knowledge and attitudes: A cross-sectional study among Saudi consumers from food trucks owned by productive families. *International Journal of Environmental Research and Public Health*, *19*(7), 4322.

Azhar, Z., Anis, A., Putra, H. S., & Juneldi, J. (2021, June). Analysis of community participation on sustainable development planning of Geopark Silokek tourism at Sijunjung Regency. In *Sixth Padang International Conference on Economics Education, Economics, Business and Management, Accounting and Entrepreneurship (PICEEBA 2020)* (pp. 201–205). Atlantis Press.

Ajzen, I. (2020). The theory of planned behavior: Frequently asked questions. *Human Behavior and Emerging Technologies*, *2*(4), 314–324.

Baker, D. M. A., & Unni, R. (2022). Understanding residents' opinions and support toward sustainable tourism development in the Caribbean: The case of Saint Kitts and Nevis. *The Coastal Business Journal*, *18*(1), 1.

Barclay, M. J., Smith, C. W., & Watts, R. L. (1995). The determinants of corporate leverage and dividend policies. *Journal of Applied Corporate Finance*, *7*(4), 4–19.

Batat, W. (2021). The role of luxury gastronomy in culinary tourism: An ethnographic study of Michelin-Starred restaurants in France. *International Journal of Tourism Research*, *23*(2), 150–163.

Berno, T., Rajalingam, G., Miranda, A. I., & Ximenes, J. (2021). Promoting sustainable tourism futures in Timor-Leste by creating synergies between food, place, and people. *Journal of Sustainable Tourism*, 1–15.

Bruley, E., Locatelli, B., & Lavorel, S. (2021). Nature's contributions to people: Coproducing quality of life from multifunctional landscapes. *Ecology and Society*, *26*.

Chae, H., Kim, S., Lee, J., & Park, K. (2020). Impact of product characteristics of limited-edition shoes on perceived value, brand trust, and purchase intention; focused on the scarcity message frequency. *Journal of Business Research*, *120*, 398–406.

Chatterjee, S., & Suklabaidya, P. (2021). Gap between perception and satisfaction: Exploring food tourism in New York. *International Journal of Hospitality and Tourism Systems*, *14*(1), 90–97.

Compeau, D., Higgins, C. A., & Huff, S. (1999). Social cognitive theory and individual reactions to computing technology: A longitudinal study. *MIS Quarterly*, 145–158.

Damastuti, E., Adventini, N., Oginawati, K., Syahfitri, W. Y. N., Kurniawati, S., & Santoso, M. (2021). Preliminary assessment of heavy metals in street vended foods in the surrounding of elementary school area in Bandung city, Indonesia. In *IOP Conference series: Materials science and engineering* (Vol. 1011, No. 1, p. 012069). IOP Publishing.

Dancausa Millán, M. G., Millán Vázquez de la Torre, M. G., & Hernández Rojas, R. (2021). Analysis of the demand for gastronomic tourism in Andalusia (Spain). *PLoS One*, *16*(2), e0246377.

Ezeh, P. C., & Nkamnebe, A. D. (2022). Determinates of consumer patronage of a street food vendor in Nigeria. *Journal of Foodservice Business Research*, 1–23.

Florek, M., & Gazda, J. (2021). Traditional food products – between place marketing, economic importance and sustainable development. *Sustainability*, *13*(3), 1277.

Fornell, C., & Larcker, D. F. (1981). Evaluating structural equation models with unobservable variables and measurement error. *Journal of Marketing Research*, *18*(1), 39–50.

Frempong, C. S., Charles-Unadike, V. O., Anaman-Torgbor, J. A., & Tarkang, E. E. (2022). Correlates of intention to practice good food hygiene among street food-vendors in Kadjebi District, Ghana. *International Journal of Health Promotion and Education*, 1–13.

Fritsche, F., Brunori, G., Chiaramonti, D., Galanakis, C., Matthews, R., & Panoutsou, C. (2021). *Future transitions for the Bioeconomy towards Sustainable Development and a Climate-Neutral Economy-Bioeconomy opportunities for a green recovery and enhanced system resilience*.

Gerritsen, R. (2020). Food walks and street doctors: Health and culinary nostalgia in a south Indian city. In *Food identities at home and on the move* (pp. 94–107). Routledge.

Gryech, I., Ben-Aboud, Y., Guermah, B., Sbihi, N., Ghogho, M., & Kobbane, A. (2020). MoreAir: A low-cost urban air pollution monitoring system. *Sensors*, *20*(4), 998.

Guadalupe-Moyano, V., Villagómez-Buele, C., Carvache-Franco, M., Carvache-Franco, W., & Ramón-Casal, T. (2022). Evaluation of the hygienic quality of the gastronomic offer of a coastal tourist destination: A study in San Pablo, Ecuador. *Foods*, *11*(6), 813.

Gupta, K., & Mohta, S. (2022). Future of food tourism in India: A psychographic overview. In *Event tourism in Asian countries* (pp. 171–182). Apple Academic Press.

Hall, R. (2017). Southeast Asia: New views of the geology of the Malay Archipelago. *Annual Review of Earth and Planetary Sciences*, *45*, 331–358.

Hanser, A. (2021). Good food in the city: How cultural ideas about food shape street vending regulation. *International Journal of Urban and Regional Research*, *45*(3), 519–534.

Hashim, R., & Jamaluddin, M. R. (2022). Co-creation of destination image and food tourism of the emerging economies. In *Gastronomy, hospitality, and the future of the restaurant industry: Post-COVID-19 perspectives* (pp. 244–257). IGI Global.

Hernández-Rojas, R. D., & Huete Alcocer, N. (2021). The role of traditional restaurants in tourist destination loyalty. *PLoS One*, *16*(6), e0253088.

Henseler, J., Dijkstra, T., Sarstedt, M., Ringle, C., Diamantopoulos, A., Straub, D., . . . Calantone, R. (2014). Common beliefs and reality about partial least squares: Comments on Rönkkö and Evermann. *Organizational Research Methods*, *17*(2), 182–209.

Henseler, J., Ringle, C. M., & Sarstedt, M. (2015). A new criterion for assessing discriminant validity in variance-based structural equation modeling. *Journal of the Academy of Marketing Science*, *43*(1), 115–135.

Hilmi, M. (2020). Street food vendors' entrepreneurial marketing characteristics and practices from 12 countries: What lessons can be learnt for improving food marketing in BOP/Subsistence marketplaces. *Middle East Journal of Agriculture Research*, 321–348.

Hsu, F. C., & Scott, N. (2020). Food experience, place attachment, destination image and the role of food-related personality traits. *Journal of Hospitality and Tourism Management*, *44*, 79–87.

Huynh-Van, B., Vuong-Thao, V., Huynh-Thi-Thanh, T., Dang-Xuan, S., Huynh-Van, T., Tran-To, L., . . . Nguyen-Viet, H. (2022). Factors associated with food safety compliance among street food vendors in Can Tho city, Vietnam: Implications for intervention activity design and implementation. *BMC Public Health*, *22*(1), 1–11.

Ince, E., & Ozturen, A. (2021). Strengths, weaknesses, and challenges of municipalities in North Cyprus aspiring to be a sustainable Cittaslow tourism destination. *Worldwide Hospitality and Tourism Themes*, *13*(4), 528–534.

Ismail, N. A. B. (2020). *A comparison of how two homestays (Kampong Beng Homestay and Gopeng Homestay) safeguard traditional Malay food (TMF) as a means of promoting cultural-based food tourism in Perak, Malaysia* (Doctoral dissertation, Newcastle University).

Jeaheng, Y., & Han, H. (2020). Thai street food in the fast-growing global food tourism industry: Preference and behaviors of food tourists. *Journal of Hospitality and Tourism Management*, *45*, 641–655.

Knollenberg, W., Duffy, L. N., Kline, C., & Kim, G. (2020). Creating competitive advantage for food tourism destinations through food and beverage experiences. *Tourism Planning & Development*, 1–19.

Kumbara, V. B., & Afuan, M. (2020). The impact of product innovation and market orientation on the competitive advantage of souvenirs in Padang. *Journal Ekobistek*, *9*(1).

Laland, K., & Seed, A. (2021). Understanding human cognitive uniqueness. *Annual Review of Psychology*, *72*, 689–716.

Laohaviraphap, T., & Wetchasart, P. (2021). The effect of risk and image on repurchase intention of street food in the Covid-19 era. *International Journal of Social Science Research*, *3*(2), 34–43.

Lee, D., Chen, T., & Chan, W. (2021). Perceptions of Gen Z tourists on street food in Hong Kong. In *Generation Z marketing and management in tourism and hospitality* (pp. 249–279). Palgrave Macmillan.

Lee, S. T., & Kim, H. S. (2021). Food fight: Gastrodiplomacy and nation branding in Singapore's UNESCO bid to recognize Hawker culture. *Place Branding and Public Diplomacy*, *17*(2), 205–217.

Li, S., Liu, X. X., Cai, S., & Scott, N. (2021). Vegan tours in China: Motivation and benefits. *International Journal of Tourism Research*, *23*(2), 238–252.

Lugosi, P., Robinson, R. N., Walters, G., & Donaghy, S. (2020). Managing experience co-creation practices: Direct and indirect inducement in pop-up food tourism events. *Tourism Management Perspectives*, *35*, 100702.

Maglumtong, M., & Fukushima, S. (2022). Inclusive development for street food vendors in modernizing Bangkok, Thailand an evaluation of the street food program of the Government Savings Bank. *Urban and Regional Planning Review*, *9*, 42–62.

May, R. Y. Y., Aziz, K. A., Latif, R. A., Latip, M. S. A., Kwan, T. C., & Kadir, M. A. A. (2021). The success factors affecting street food stalls for gastronomic tourism competitiveness: A case of Petaling Jaya Old Town. *International Journal of Early Childhood Special Education (INT-JECSE)*, *13*(1), 241–256.

Mitgosoom, Y., & Ashton, A. S. (2019). Thai street food and brand image development: An investigation from tourists' perspective. *Journal of Tourism, Hospitality and Culinary Arts*, *11*(2), 69–82.

Moghavvemi, S., Woosnam, K. M., Hamzah, A., & Hassani, A. (2021). Considering residents' personality and community factors in explaining satisfaction with tourism and support for tourism development. *Tourism Planning & Development*, *18*(3), 267–293.

Mohamad, N., S. Palan, D., Roslan, M. A., & Nasron, N. A. (2022). Predictors of behavioral intention among tourist: The case of revisiting street food spots in Penang, Malaysia. *Journal of Foodservice Business Research*, *25*(4), 475–497.

Mortazavi, R. (2021). The relationship between visitor satisfaction, expectation, and spending in a sport event. *European Research on Management and Business Economics*, *27*(1), 100132.

Mudunkotuwa, M. D. M., & Arachchi, R. S. S. W. (2020). The impacts of tourists' perceived risk on attitude and behavioral intention towards street food: A case study of central colombo. *International Journal of Tourism Research*, *4*(2).

Muis, A., Luneto, R., & Safitri, D. (2021, March). *Micro, small, medium enterprises business empowerment to creative-halal industry: Street food in Bangkok*. Proceedings of the

5th International Conference on Indonesian Social and Political Enquiries, ICISPE 2020, 9–10 October 2020, Semarang, Indonesia.

Mwove, J., Imathiu, S., Orina, I., & Karanja, P. (2020). Food safety knowledge and practices of street food vendors in selected locations within Kiambu County, Kenya. *African Journal of Food Science*, *14*(6), 174–185.

Nazir, M. U., Yasin, I., Tat, H. H., Khalique, M., & Mehmood, S. A. (2021). The influence of international tourists' destination image of Pakistan on behavioural intention: The roles of travel experience and media exposure. *International Journal of Hospitality & Tourism Administration*, 1–25.

Naruetharadhol, P., & Gebsombut, N. (2020). A bibliometric analysis of food tourism studies in Southeast Asia. *Cogent Business & Management*, *7*(1), 1733829.

Nofrizal, N. (2022). The key to success increasing customer satisfaction of SMEs product souvenir food. *Cogent Business & Management*, *9*(1), 2084897.

Okojie, P. W., & Isah, E. C. (2019). Food hygiene knowledge and practices of street food vendors in Benin City, Nigeria. *International Journal of Consumer Studies*, *43*(6), 528–535.

Okumus, B., Mehraliyev, F., Ma, F., & Köseoglu, M. A. (2021). Intellectual connections in food tourism literature: A co-citation approach. *International Journal of Tourism Research*, *23*(2), 220–237.

Palonen, J. (2022). *Developing a customer value proposition for the street food segment: Case Teuvan Keitintehdas*.

Park, E., Muangasame, K., & Kim, S. (2021). "We and our stories": Constructing food experiences in a UNESCO gastronomy city. *Tourism Geographies*, 1–22.

Perdana, M. R., & Wibowo, L. A. (2021). Culinary experience toward behavioural intention (survey of consumer fusion food on street food in Bandung city). In *Promoting creative tourism: Current issues in tourism research* (pp. 446–452). Routledge.

Pilato, M., Platania, M., & Séraphin, H. (2021). Connecting locals and visitors: The case of street food. In *Tourism in the Mediterranean Sea*. Emerald Publishing Limited.

Poltanee, I., & Thavornsil, T. (2022). Destination image of street food on China town an impact on tourists' dining experiences. *Central Asia and the Caucasus*, *23*(1), 191–201. https://doi.org/10.37178/ca-c.23.1.017

Praesri, S., Meekun, K., Lee, T. J., & Hyun, S. S. (2022). Marketing mix factors and a business development model for street food tourism. *Journal of Hospitality and Tourism Management*, *52*, 123–127.

Prasongthan, S., & Silpsrikul, R. (2022). The structural equation model of behavioural intentions toward gastronomy tourism in Thailand. *Kasetsart Journal of Social Sciences*, *43*(2), 425–432.

Rakha, A., Fatima, M., Bano, Y., Khan, M. A., Chaudhary, N., & Aadil, R. M. (2022). Safety and quality perspective of street vended foods in developing countries. *Food Control*, 109001.

Reinartz, W., Haenlein, M., & Henseler, J. (2009). An empirical comparison of the efficacy of covariance-based and variance-based SEM. *International Journal of Research in Marketing*, *26*(4), 332–344.

Ringle, C. M., Sarstedt, M., & Mooi, E. A. (2010). Response-based segmentation using finite mixture partial least squares: Theoretical foundations and an application to American customer satisfaction index data. *Annals of Information Systems*, *8*, 19–49.

Rocha, A., Azevedo, D., & Silva, S. (2021, May). The impact of tourism in the restaurant business: A case study in the City of Oporto. In *ICTR 2021 4th international conference on tourism research* (p. 460). Academic Conferences International.

Rohith, S. (2021). *An investigation into the hygiene practices and food safety of street vendors outside pension pay-out points in urban poor communities in the City of Cape Town* (Doctoral dissertation, Stellenbosch, Stellenbosch University).

Sangkaew, N., & Zhu, H. (2022). Understanding tourists' experiences at local markets in Phuket: An analysis of TripAdvisor reviews. *Journal of Quality Assurance in Hospitality & Tourism, 23*(1), 89–114.

Santana, G. R., & Amparo-Santos, L. (2019). Institutionalization of street food in the modern urban space: Conformations, tensions, and conflicts. *Journal of Food and Nutrition Research, 7*(6), 465–475.

Santoso, H. N., Widyawati, C., & Adityaji, R. (2021). The role of the promotional video "Wonderful Indonesia: A visual journey" on brand image and tourist visiting interest to Wakatobi. *Journal of Tourism, Culinary and Entrepreneurship (JTCE), 1*(1), 48–60.

Seo, K. H., & Lee, J. H. (2021). Understanding risk perception toward food safety in street food: The relationships among service quality, values, and repurchase intention. *International Journal of Environmental Research and Public Health, 18*(13), 6826.

Sgroi, F., Modica, F., & Fusté-Forné, F. (2022). Street food in Palermo: Traditions and market perspectives. *International Journal of Gastronomy and Food Science, 27*, 100482.

Shen, L., & Springer, T. M. (2022). The odd one out? The impact of property uniqueness on selling time and selling price. *Journal of Housing Research*, 1–21.

Solon, J. J. C. (2022). Food safety practices among street food vendors in the Twin Cities of Zamboanga Del Norte. *International Journal of Multidisciplinary: Applied Business and Education Research, 3*(7), 1359–1368.

Soon, J. M. (2019). Rapid Food Hygiene Inspection Tool (RFHiT) to assess hygiene conformance index (CI) of street food vendors. *LWT, 113*, 108304.

Sousa, S., de Morais, I. L., Albuquerque, G., Gelormini, M., Casal, S., Pinho, O., . . . Padrão, P. (2022). Patterns of street food purchase in cities from Central Asia. *Frontiers in Nutrition, 9*.

Thuan, T. V., Chi, N. T. P., & Trung, N. T. (2018). Factors affecting international tourists' satisfaction of street food in Ho Chi Minh city. *Journal of International Economics and Management*, (111), 87–100.

Thyne, M., Woosnam, K. M., Watkins, L., & Ribeiro, M. A. (2022). Social distance between residents and tourists explained by residents' attitudes concerning tourism. *Journal of Travel Research, 61*(1), 150–169.

Tymoshchuk, O., Almeida, A. M., Pedro, L., Antunes, M. J., Ramos, F., Oliveira, E., & Carvalho, D. (2021). Digital technologies as tools to promote tourism and territorial development: Design of a mobile application for community-led initiatives. In *Impact of new media in tourism* (pp. 268–291). IGI Global.

Wang, M., Jiang, J., Xu, S., & Guo, Y. (2021). Community participation and residents' support for tourism development in ancient villages: The mediating role of perceptions of conflicts in the tourism community. *Sustainability, 13*(5), 2455.

Wei, C. K., Ying, G. J., & Tze, H. K. (2022). Chinese domestic visitors' attitudes towards street food in Zhongshan Road Night Market, Nanning, China. *International Journal of Early Childhood, 14*(3).

Wiatrowski, M., Czarniecka-Skubina, E., & Trafiałek, J. (2021). Consumer eating behavior and opinions about the food safety of street Food in Poland. *Nutrients, 13*(2), 594.

Yusof, M. F. M., & Simpong, D. B. (2021). Exploring the potential of Pondok Institutions as an Islamic spiritual tourism product – the case of Malaysia. In *The importance of new technologies and entrepreneurship in business development: In the context of economic diversity in developing countries: The impact of new technologies and entrepreneurship on business development* (pp. 722–730). Springer International Publishing.

Zhang, T., Chen, J., & Hu, B. (2019). Authenticity, quality, and loyalty: Local food and sustainable tourism experience. *Sustainability*, *11*(12), 3437.

Zhu, M., Huang, D., Hu, X. J., Tong, W. H., Han, B. L., Tian, J. P., & Luo, H. B. (2020). Application of hyperspectral technology in detection of agricultural products and food: A Review. *Food Science & Nutrition*, *8*(10), 5206–5214.

Zopiatis, A., & Pericleous, K. (2021). Profiling the package traveler: An expenditure-based segmentation endeavor. *Journal of Destination Marketing & Management*, *21*, 100636.

11

KNOWLEDGE LEGACY OF THE 2010 FIFA WORLD CUP IN SOUTH AFRICA

Willem JL Coetzee

Introduction

"The 2010 World Cup will be organised by . . . South Africa!" FIFA president Sepp Blatter announced on the 15th of May 2004. Former president Nelson Mandela then climbed onto the stage, lifted the World Cup trophy, and said: "I feel like a young man of 15." In response to this announcement, the South African bid Chief Mr Irvin Khoza: "This is for Africa. This is for African renewal" and "Ke Nako" – "Celebrate Africa's Humanity" (Blackshaw & Coetzee, 2020, p. 369). These images and pronouncements sparked the immediate attention of some researchers, and academic journals started to accept publications related to the event 5 years before the 2010 FIFA World Cup (2010-WC). This mega-event in Africa provided me (as an Early Career Researcher in 2004) and other academics on the continent with the opportunity to create knowledge relevant to Africa. The research skills and knowledge gained should be regarded as the knowledge legacy of the event. Although workshops were organized to facilitate a 2010 research agenda, it was never formalized due to the timeframes and the lack of political motivation to allocate the necessary resources to implement such a coordinated research agenda (Swart & Bob, 2012). For this reason, 2010-WC research took place on an ad hoc basis, and it is, therefore, appropriate to reflect on the research related to the first FIFA World Cup hosted in Africa. The objective of this paper is to inspect the knowledge legacy of this event since that day in May 2004. There appears to be a lack of studies that present a systematic overview of journal publications related to the 2010 FIFA World Cup (2010-WC). The knowledge legacy of the 2010-WC is therefore still relatively unknown.

In the context of mega-events in sport, Preuss (2007, p. 211) proposes the following definition of legacy: "irrespective of the time of production and space, legacy can be planned and unplanned, positive and negative, tangible and intangible

DOI: 10.4324/9781032633374-13

structures created for an by a sporting event that remain longer than the event itself." Chappelet and Junod (2006, p. 85) refer to the term "collective memory" and distinguish in this regard between five types of legacies, namely sporting; urban; infrastructural; economic and social legacies. An essential part of the social legacy of mega-events is the change that is brought about in residents' perceptions of the host city or region, as well as the memories of the mega-event – including the skills and experiences that residents gained through their direct or indirect involvement in the event. Within this realm of social legacies, we can place the knowledge legacy of an event. In Chappelet and Junod's review of legacies, nothing is explicitly mentioned about the knowledge legacy of these events. Even after the 2010-WC, Cornelissen et al. (2011), summarized the positive and negative legacies of the 2010-WC legacy. In their 2012 publication, the same authors refer to some research projects, but in neither of these publications was the knowledge legacy of the 2010-WC clearly articulated. Knowledge legacy could be seen as part of knowledge management, where knowledge management refers to the creation, identification, acquisition, application, storage, and transfer of knowledge. These activities were identified in 2009 by Heisig when he aimed to harmonize knowledge management by comparing 160 knowledge management frameworks. In the end, he placed these six activities as the crux of knowledge management. Various other knowledge management activities and frameworks exist, and debates surrounding information versus knowledge are well-defined. According to Davenport and Prusak (1998), knowledge may be understood as experiences, contextual comprehension, value-added information, or insights based on frameworks of understanding that reside in the minds of individuals or groups. Knowledge may be present in the form of explicit or tacit knowledge. Detailed knowledge is more easily articulated, written, or codified, while tacit knowledge is rather inarticulate, developed with experience, and deepened through problem-solving activities (Polanyi, 1967). As evident from this chapter, the interdisciplinary field of event and sports management knowledge is vast and diverse, and a proper research agenda should be an essential part of any mega-event project plan.

The chapter is structured into the following four sections: the next section defines the legacy of sporting events; the second section describes the methods used in this article; the third section presents the results of this review regarding pre-, during, and post-2010-WC research; and the final section, further findings and their implications are discussed outlining future research possibilities.

The legacy of sporting events

Harper (2001) defined legacy as property left by will, while Cashman (2005) categorized event legacies in terms of economic; built and physical environment; information and education; public life, politics, and culture; sport; symbols, memory, and history. As mentioned earlier, a similar classification of legacies was proposed by Chappelet and Junod (2006, p. 85); in this regard, they also refer to the term "collective memory" and distinguish between the five types of legacies mentioned earlier. Preuss (2007, p. 211) recommends using the term "legacy-cube" and defines mega-event "legacy" as: "irrespective of the time of production and space, legacy can be planned and unplanned, positive and negative, tangible and intangible structures created for and by a sports event that remains longer than the event itself." This notion of a legacy-cube relates to why nations often invest social, environmental and economic resources to host mega-sports events.

Over the past two decades, nations worldwide have competed to host mega-sporting events because countries (and increasingly cities) perceive such events as potential re-imaging opportunities (Hiller, 2006; Smith, 2005). They also believe such events are likely to boost their image as tourism destinations Hall & Page, 2009), that this will promote investment, and that the nation is likely to benefit from the associated positive economic impacts (Briedenhann, 2011). Especially for developing nations, mega-events are integral in redefining a country's global status and socio-political composition (Cornelissen, 2008). These impacts are known as the event legacy and include the opportunity to increase tourism value; brand the hosting country as a safe tourist destination; advance the social legacy of the country by creating employment, equity and growth; advance urban planning and sports infrastructure; create more local business opportunities and a renewed community spirit; and bring about additional know-how and a positive emotional experience (Preuss, 2007). For many African researchers, the 2010-WC event acted as a platform to establish their academic careers.

According to Getz and Page (2015), there has been an over-emphasis on knowledge creation in the realm of economic impacts of events. They further state that interest in events continues to expand within all closely related and many foundation disciplines, giving rise to new theoretical perspectives and useful methodologies. They say that each core proposition of event tourism should give rise to collaborative theory development. A similar possibility was posed by Gibson (1998) in her critical view that an interdisciplinary approach is necessary when formulating a research agenda. The successful hosting of a mega event, and more importantly, achieving a positive legacy, is grounded in the acquisition, production and dissemination of information and knowledge and should not be overlooked. However, research requires sufficient planning, high levels of interdisciplinary cooperation, commitment and resources (Swart & Bob, 2012).

Event and tourism studies are mostly multidisciplinary, drawing theory, knowledge, methodologies, and methods from many established disciplines (Getz & Page, 2015). Authors such as Getz and Page (2015) refer to this multidisciplinary nature as important for knowledge creation; they state that this field has largely been ad hoc and fragmented among diverse interest groups. In this regard, they specifically suggest that research on the process and research related to actors in knowledge creation for event tourism need to be more present. It will be evident from this article that knowledge creation pre-, during, and after the 2010-WC is one of the legacies of this mega-event.

Significantly, a principal issue for all concerned with the organization of the 2010-WC was that of legacy (Allen et al., 2013), and an essential aspect of the legacy discussed the long-term impact (both financial and social) of the event on the existing national brand of South Africa. After the 2010-WC, Cornelissen et al. (2011) tabled a range of material, spatial and symbolic legacies, both positive and negative, that a host city and its wider region could experience. Unfortunately, in their discussion (and simplified model) of legacy impacts, nothing is mentioned about knowledge creation and the knowledge legacy of the 2010-WC.

Methodology

The premise of this article is from Kim et al. (2013), who emphasized that academic journals play an important role in the work undertaken by researchers because they not only produce, disseminate, and exchange knowledge, but articles also provide a means of evaluating research. These authors elaborate that a primary goal of analysing the topics examined in studies is to identify research trends and future direction. The same can be said for journal articles related to a particular event, such as the 2010-WC, since these articles disseminate and exchange academic knowledge. By analysing the content of these articles, the knowledge legacy of this event may be better defined, but the author fully recognizes the importance of other seminal works such as the "International Handbook on the Economics of mega sporting events (Maennig & Zimbalist, 2012)," conference papers, dissertations, industry reports, and other publications in the popular media, that were all excluded from the analysis, purely on the basis that for the purpose of this article the researcher only refers to peer-reviewed journal articles.

Formica (1998) was one of the first researchers in the sphere of event management to evaluate research outputs in event studies by conducting content analyses on research articles published on special events and festivals. Formica classified them according to the event's geographical location, the authors' location, the journals that contributed the most to the field, the nature of the articles, the methodology and statistical techniques used, and the functional area under investigation. For the present chapter, the author decided to use a similar approach to a combination of systematic literature review and content analysis.

A systematic literature review was used as a starting point to identify academic publications, and content analysis was used as a scientific technique to analyze the publications. Klassen et al. (1998) define the systematic review as follows: "A systematic review is a review in which there is a comprehensive search for relevant studies on a specific topic, and those identified are then appraised and synthesised according to a pre-determined explicit method" (Klassen et al., 1998, p. 701). In the context of sports management, Weed (2006) states that this synthesis method is widespread in other fields and that systematic reviews differ from traditional narrative reviews in that they provide objective, replicable, systematic, and comprehensive coverage of a defined area.

Data source: As a starting point for the current research project, the author searched and identified articles related to the 2010-WC. The search was narrowed by looking for articles published in English that appeared in peer-reviewed journals containing words (or a combination of these words) such as South Africa(n); FIFA; World Cup™; 2010; soccer; and football. A dataset of articles was compiled from bibliographic databases such as ResearchGate, Google, ScholarGoogle, ScienceDirect, EBSCOhost, and Scopus. The spreadsheet was cleaned and simplified by deleting entries such as book chapters, research notes, technical reports, papers in conference proceedings, master and doctorate dissertations, and duplicated article entries. These entries were only deleted to frame this particular study and by no means intended to disregard the knowledge contribution of other research outputs. The final spreadsheet comprises 185 articles published over 11 years (2004–2015). Although "legacy" is measured in the long term, the timeframe was only selected to frame the sample of this project. Otherwise, the collection will be forever ongoing (as, indeed, it should be with legacy projects).

The 185 full papers were downloaded, and the author set out to analyze the content of the articles by reading the article titles and abstracts to verify that each item refers to the 2010-WC. Content analysis is described as making replicable and valid inferences from data in its context (Krippendorff, 2012) via reflexive analysis of documents (Altheide, 1987) and, in the case of the present article, of the content of 2010-WC-related research articles. To ensure objectivity and consistency of analysis, the researcher did all aspects of the research analysis. After the analysis stage, the various researched areas were categorized into the following groups: the title of the article; date of publication; the name of the journal; keyword(s) of the article, the topic covered in the article, the name of the author(s); country of origin of the author(s) at the time of publication; and the methodology.

Data analysis: In addition to the systematic review, text analysis was conducted through Leximancer text analytics software (4.0 edition) to confirm the themes identified in the study. According to Pearce and Wu (2016), Leximancer is a relatively new method for transforming lexical co-occurrence information from natural language into semantic patterns. This software allows the researcher to determine the main themes within a text, highlight how topics relate to each other and indicate which source files contain particular topics (Angus et al., 2013). Leximancer generates concept maps and illustrates the relationships among words, concepts and themes. It is important to note that not all the identified concepts are labeled in the concept maps. Generic concepts, such as abstract, research methodology, results, bibliography, and page number (p.) in research articles are not labeled in the maps that we used for analysis in this chapter.

Results

A historical review of the 2010-WC publications

For the purpose of the present chapter and with a view to present information in a logical flow, a historical review of publication pre-, during, and post-2010-WC is presented. This particular classification is to assist future research teams when planning their research agendas before, during, and after a mega-event.

Pre-2010 World Cup publications

Articles published before the 2010-WC focused on expectations, predictions and politics associated with this event. The Leximancer analysis showed that the major research themes during this time gravitated around hosting the event in South Africa, government involvement, economic benefits, tourism, and soccer. As mentioned before, the unidentified circles within the maps are general journal-related themes, such as abstract, bibliography, literature review, and page numbers.

Eleven years before the 2010-WC, and only days after the announcement by FIFA that South Africa would be the host country, the first two journal articles were published. One focused on the "way to prosperity" for a developing nation (Matheson & Baade, 2004), and in the second article called "It is Africa's turn!" the narratives and legitimations surrounding the Moroccan and South African bids for the 2006 and 2010 FIFA finals were explored (Cornelissen, 2004). In a follow-up article, Cornelissen and Swart (2006) considered "the challenges of making good on an African promise," and in 2007 Maennig and Du Plessis (2007) focused on economic perspectives and policy challenges informed by experiences during the 2006 World Cup in Germany.

In the same year, the first special edition of an academic journal focusing on the 2010-WC was published by *Politikon*. As the official publication of the South African Association of Political Studies, it was appropriate that the editor called this special edition "Crafting legacies: The changing political economy of Global Sport and the 2010 FIFA World Cup" (Cornelissen, 2007). In this special edition, authors focused on issues such as the politics of mega-events (Black, 2007), sport and national identity (Kersting, 2007), migration during a mega-event (Cornelissen & Solberg, 2007), the political economy of mega-stadiums and mega-projects (Alegi, 2007; Van Der Westhuizen, 2007), perceptions of crime at a host region (Donaldson & Ferreira, 2007), and a sport tourism strategy beyond 2010 (Swart & Bob, 2007).

Two years before "kick-off," most of the publications focused on the sustainability of the stadia (Alegi, 2008), as well as other social aspects, such as the economic value of the event as a tool for poverty reduction (Bohlmann & Heerden, 2008; Pillay & Bass, 2008; Saayman & Rossouw, 2008), the socio-economic impacts associated with the event (Donaldson et al., 2008; Nadvi, 2008), and in particular the aspects around nation-building during the run-up to the event (Labuschagne, 2008). It was also in this year that the first article on health-related research was published when Yancey Ii et al. (2008) discussed an integrated public health and medical system for the 2010-WC.

In 2009, a second special issue, this time from the journal *Urban Forum* was published, focusing on critical issues faced by some of the host cities. These issues included safety and security in an urban destination, a crime strategy for a host city, and the perceived economic benefits for B&B operators in one of the provinces (Darkey & Horn, 2009). In this special edition, there were also articles related to hosting cities in terms of sex tourism (Bird & Donaldson, 2009), the Cape Town Stadium (Bob & Swart, 2009), the beautification of the N2 highway toward Cape Town (Newton, 2009), and several urban development ambitions of Durban (Maennig & du Plessis, 2009).

Publications during 2010

According to the Leximancer analysis, the research themes in 2010 centered around the World Cup itself, the host continent, national identity, and the people involved in hosting the event, especially the workforce and the role of women.

During 2010, a relatively high number of articles about health sciences were published. The number of publications related to the 2010-WC and health sciences might be ascribed to the relatively transient nature of articles in health science journals and, therefore, a quick turnaround time from the submission date to the date of publication. For example, Jazaerly and Giorgadze (2010) published a one-page article in the *Journal of Surgical Pathology* related to lymph nodes, and both Fieggen (2010) and Guinto Balanzar (2010) published one-page article related to neurosurgery and the 2010-WC in the *Journal of World Neurosurgery*. Other examples of health-related articles are those related to travel health (Leggat et al., 2010), vaccines (Zuckerman et al., 2010), rabies in Africa (Malerczyk et al., 2010), and health risks of travelers (Mendelson et al., 2010) as well as public health and sex workers (Richter et al., 2010).

Apart from publications in the sphere of health sciences, a significant number of articles were also published in journalism and sociology. Vincent et al. (2010) explored narratives in English newspapers during the World Cup, while Berger (2010) studied visual images of the World Cup. Ginsberg (2010) and (Aranke & Zoller, 2010) published articles related to neoliberalism and politics in Africa, and Ngonyama (2010) referred to the critical voice in the journal *Soccer and Society*. Swart et al. (2010) focused on media- and crime-related issues, and Giulianotti and Klauser (2010), as well as Roberts (2010), studied security governance, policing, and mega-sports events.

In 2010, the journal *Agenda* published a special edition in which most of the articles focused on gender issues. For example, Bob and Swart (2010a) looked at the experiences of women in fan parks, Hicks (2010) wrote a gender-based response to the 2010-WC, while Pillay and Salo (2010) explored the relationship between gender, politics and sport.

Another special edition during this year was from the journal *Alternation* edited by Bob and Swart (2010b). This special edition presented research, mostly by the editorial team, on a variety of issues. Smit (2010) wrote an article on Dennis Brutus, while Bob and Swart (2010a, 2010b) addressed social legacies; Chain and Swart

(2010) presented residents' perceptions of the World Cup, while Tichaawa and Swart (2010) focused on fans from Cameroon; McKenna and Bob (2010) addressed business perceptions; and Swart et al. (2010) perceptions in the media. The other authors in this special issue addressed the political economy (Cornelissen & Maennig, 2010); living in close proximity to the stadium in Durban (Bassa & Jaggernath, 2010); other sports events and tourism in Durban (Maharaj, 2010); the environmental impacts of a mega-event such as the 2010-WC (Ahmed & Pretorius, 2010) and service delivery (Cronjé et al., 2010).

Publication after 2010

According to the Leximancer analysis, within the first year after the 2010-WC, 41 articles related to the event were published. Within the social sciences, researchers interrogated linkages between the 2010-WC and art (Dubin, 2011), the media (Hammett, 2011) and music (Hammond, 2011), as well as other broader social impacts (Van Blerk, 2011) of the event. Some studies explored more specific impacts such as sex workers (Delva et al., 2011) and human trafficking (Roelofse, 2011). Apart from these issues, some researchers studied security in South Africa during the World Cup (Cornelissen, 2011a); smoke-free laws in Guatemala and the effect of these on 2010-WC enjoyment while watching in public places (GTQ, 2011). Some studies engaged with the use of technology in strategic policing within the South African Development Community (Sinclair & Matlala, 2011) as well as the politics of the event (Harris, 2011; Masemola & Chaka, 2011; Ndlovu-Gatsheni, 2011).

Researchers in the economic sciences studied the role of advertising (Conradie, 2011), marketing (Cubizolles, 2011) and entrepreneurship (Madichie, 2011) during the mega-event as well as the economics (du Plessis & Maennig, 2011) and economical recovery (Ferreira, 2011) due to the event. Within the realm of tourism authors, a number of publications (Bresler, 2011; Lepp & Gibson, 2011; Tomlinson et al., 2011; Vincent & Hill, 2011) studied issues related to identity, the hosting of mega-events and risks associated with the event, while others looked at legacy (Cornelissen, 2011b; Cornelissen et al., 2011; van der Westhuizen & Swart, 2011), community expectations and perceptions (Bob & Majola, 2011), and expectations from a youth development perspective (Saurombe & Saurome, 2011).

Other research topics included issues such as urban development (Haferburg, 2011), the environment (Death, 2011), and project management of construction projects (Baloyi & Bekker, 2011). A large number of publications appeared in the field of health sciences, for example, HIV and public health (Beres, 2011), the impact of acoustic noise and vuvuzelas (Cunefare et al., 2011); hearing loss (Ramma, 2011); blood exposure (Furin, 2011); the medical assessment of referees (Bizzini et al., 2011), injuries and illnesses of football players (Dvorak et al., 2011), and the health of volunteers (Muxel et al., 2011). Lames and Kolbinger (2011) reported on performance analysis of clubs in Europe, while Duk et al. (2011) studied the endurance and speed of players.

As was the case in 2011, a large number of articles (n = 26) were published in 2012. Within the social sciences, the topics ranged from the lexicon of the sport of journalistic chronicles (Almaraz & Garcia, 2012) and television broadcasting (Lopes et al., 2012) to the Cape Town stadium in the media (Chuma, 2012) and a personal reflection in the *Drama Review* on the empathy, optimism, and beautiful play of the 2010-WC (Larlham, 2012). In health sciences, a number of articles were published related to the 2010-WC and psychology (Jones et al., 2012; Koortzen & Oosthuizen, 2012; Oosthuizen & Koortzen, 2012; Solari & Solomons, 2012) as well as the 2010-WC and public health (M. Richter et al., 2012), medication (Tscholl & Dvorak, 2012) and testosterone (van der Meij et al., 2012).

Authors also addressed the impact of the 2010-WC on society in a number of articles in the fields of sociology (Cornelissen, 2012), exploring quality of life (Rowe & Baker, 2012), political sciences such as Pan-Africanism (Bolsmann, 2012), race and African identity (Lopez, 2012), modernity and development hopes (Manzo, 2012), transformation (Brook Napier, 2012), nation-branding (Knott et al., 2012), patrimony (Cubizolles, 2012), perceptions of crime (George & Swart, 2012), and foreign policy (Carreño Lara, 2012). Two other areas of research that received attention during this time match analysis (Ademović et al., 2012), and in management sciences, reference was made to brand equity and entrepreneurship (Nicolau, 2012; Venter et al., 2012).

In 2013, 30 articles were published in a variety of journals. In health sciences, researchers reported on a large number of issues such as sports injuries (Guillodo & Saraux, 2013), depression (Jain et al., 2013), arterial stiffness (Reppel et al., 2013), hypoxia/altitude and endurance (Nassis, 2013), risk of injury (Ryynänen, Dvorak, et al., 2013), injury incidence (Ryynänen, Junge, Dvorak, Peterson, Karlsson, et al., 2013), injury recovery (Eirale et al., 2013), and admissions in emergency departments due to assaults associated with the 2010-WC (Quigg et al., 2013).

Researchers in the field of sports published articles related to game patterns (Machado et al., 2013), network analysis of games (Cotta et al., 2013), ball retention (Collet, 2013), and foul play (Ryynänen, Junge, Dvorak, Peterson, Kautiainen, et al., 2013). In tourism and sports studies, researchers studied the social value of sports (Heere et al., 2013), social responsibility and destination benefits (Walker et al., 2013), the quality of life of residents (Kaplanidou et al., 2013) and the impact on host communities (Hermann et al., 2013), co-creation by residents (Berndt et al., 2013), the host-guest relationship between residents and visitors (van der Zee & Go, 2013), the challenges of hosting a mega-event in one of the host cities (Allen, 2013), and destination competitiveness (Kruger & Heath, 2013).

Safety (Kirby et al., 2013) and security (Eisenhauer et al., 2013) were again visible issues; as were South Africa's diplomacy (e Castro, 2013) and racial unity (Duvall & Guschwan, 2013) during the 2010-WC. Within the field of management sciences, researchers studied branding legacy (Allen et al., 2013), media impact (Swart et al., 2013), and behavioral patterns (Menéndez et al., 2013). Others studied fan assessment from a psychological perspective (Gau, 2013) and identity (Oh, 2013).

In 2014, there was a notable decline in the number of publications (n = 21); but again, the topics cover various research areas. Social impacts such as gambling

(Masuku, 2014), cultural heritage (Jethro, 2014), moral panicking and human trafficking (Emser & Francis, 2014), and sex work and prostitution (Richter et al., 2014) were discussed together with other social issues such as the politics of hosting a mega-event (van Wyk, 2014).

Within the realm of sport, Chistofaro et al. (2014) wrote an article on aggression in football, while others studied the tactical behavior of players (Kempe et al., 2014), free kicks (Casal et al., 2014), regional qualification games to the run-up toward the 2010-WC (Stone & Rod, 2014) and game structure, speed and player patterns (Wallace & Norton, 2014). Several tourism-related articles were published in 2014, for example, articles addressing fandom (Alber & Ungruhe, 2014), national identity (Nicholson et al., 2014), tourism in developing countries (Peeters et al., 2014), the oversupply of hotel rooms (Ferreira & Boshoff, 2014), national flag and patriotism during the WC (Brown et al., 2014), and psychic income and social capital (Gibson et al., 2014). Other authors looked at security and surveillance (Fonio & Pisapia, 2014), the impact of the event on pediatric injury and mortality (Zroback et al., 2014), investor sentiment and market efficiency (Kaplanski & Levy, 2014), and global imaging and branding.

In 2015, only seven articles related to the 2010-WC and perhaps not surprisingly the legacy of the WC remained a talking point. Waardenburg et al. (2015) explored the local meanings of a sport and the legacies of that sport, while Molloy and Chetty (2015) addressed the legacies from a project management point of view and indicated lessons learned from the stadium program. de Almeida et al. (2015) compared South Africa to Brazil in terms of the legacies of the countries' two World Cups. Another African legacy program was that of Cameroon and Nigeria and an article by Tichaawa and Bob (2015) discussed the leveraging of mega-events beyond the host nation. The last three articles from 2015 discussed the long-term reputation of South Africa from a branding point of view (Holtzhausen & Fullerton, 2015), as well as nation branding opportunities (Knott et al., 2015) and South Africa as a destination for tourism and economic development (Giampiccoli et al., 2015).

From the journal titles, it is evident that social issues were a popular research focus area, followed by tourism and sports studies as well as health-related research. According to the SCImago Journal Ranking scale (SJR), the articles in the top ten highest ranked journals were a balanced combination of journals from different sciences. The knowledge legacy of the 2010-WC is vast and diverse, with topics ranging from tourism, sport and events to health, safety and politics. Well-known authors wrote some of the articles published in their respective fields, and some authors created new research networks. The following section discusses these authors and their networks in greater detail.

Research networks, methods and populations

In total, 332 (332) authors from 27 countries undertook WC-2010 research and published 185 articles. As the host nation, South African researchers authored or co-authored 81 articles plus 20 articles as co-authors with researchers from 11 other countries, meaning that South African research resulted in 91 articles. Researchers from the USA published 18 articles, followed by researchers from the UK with 14 articles. Other European researchers (from Germany, Spain, Portugal and Belgium), as well as some from Australia, published between four and seven articles per country. The remaining 19 of 27 countries published only 26 of the 185 articles.

Research methodologies

The analysis of research methods in the 185 articles indicated that 65 articles adopted a quantitative approach, 111 a qualitative approach, and nine articles used mixed methods. Quantitative methods of data collection included archival data such as newspapers (14 articles), surveys (30 articles), interviews (14 articles), personal reflections and observations (67 articles), and content analyses of broadcasts and other media (33 articles). Quantitative data were collected from 10 869 fans by ten different research groups. In total, 6,379 residents were interviewed by nine research groups before, during, and after the 2010-WC; 342 managers of accommodation establishments were interviewed by two research teams; 933 players by three researchers; one research group targeted 90 referees as their population; 19 destination managers in two separate studies; 852 patients by two health sciences studies; 54 project managers and engineers were interviewed by three research teams.

Discussion and conclusions

Hosting a large-scale sports tourism event produces economic, social and environmental legacies. These legacies are well documented in the literature, and host communities are using mega-events to leverage legacy programs after the event and one such event is the 2010-WC. Researchers and event organizers developed legacy frameworks and sport tourism strategies for the 2010-WC and other similar events. After the event, researchers and organizers reported on the positive legacy programs, and the organizers of the 2010-WC were applauded for the event's social, environmental, and economic benefits. Unfortunately, the research outputs were not recorded as a legacy of the 2010-WC. Therefore, this study set out to report comprehensively on the knowledge legacy of the 2010-WC.

This report has illuminated the research themes reported before, during and after the 2010-WC. Overall, the number of academic articles related to this single event is high with 185 articles dealing with the event appearing in peer-reviewed academic journals. The research themes can broadly be classified into tourism and event studies; health-related publications; a large variety of social impact issues; and, to a lesser extent, some articles addressed issues about the environmental and economic aspects. Interestingly, unlike research on previous events that has tended to overemphasize the economic value of events, the 2010-WC research related to economic aspects was outnumbered by publications related to social and health-related aspects.

The results show that the 2010-WC attracted a large number of researchers from different fields and different countries and that the knowledge legacy of the 2010-WC might be one of the greatest legacies of this event. The 2010-WC gave researchers, especially early career researchers in South Africa, the opportunity to showcase their research abilities. To reiterate what Irvin Khoza said after the announcement that South Africa will host the 2010-WC: "This is for Africa. This is for African renewal." "Ke Nako" – Celebrate Africa's Humanity – and we can add to this "Africa's knowledge legacy."

Limitations and future research

The limitations of the results of this article are that the author only explored academic journal articles and not book chapters or papers delivered and published in conference proceedings. Also, the author only used articles written in English, and articles in other local (South African) languages such as Afrikaans were excluded from the study. The author also found a comparative article in Spanish (comparing 2010-WC with the 2014 World Cup in Brazil) that was excluded from the analysis. Another limitation is that researchers are constantly producing articles, and it is possible that publications might appear in journals toward the end of 2015 that were not in the public domain at the time of data gathering. For future research, the authors would recommend a research strategy for mega-events that can be used by event organizers in future. Such a research plan or framework will ensure that the research will have a lasting legacy for sport, tourism, and event studies.

Funding for research is a global issue, and researchers are not always aware of the funding opportunities that are available for research related to mega-events. In a similar vein, the event organizers are not always aware of the potential knowledge legacy of their events, and without budgeting for research, it might be a missed opportunity for them to measure the knowledge legacy of an event. From the author's own perspective, it was noticeable that South African academics scrambled for research funding in the run-up to the 2010-WC. Often, academic institutions competed with each other for funding instead of working as one group to secure a collaborative national research fund for 2010-WC research projects. The author fully supports the statement by Swart and Bob (2012) that it was clear that there were willingness and commitment to undertake research; however, the issue of accessing the necessary resources/funding was a major challenge experienced.

Research collaborations between national institutions were therefore limited, and most South African researchers worked with researchers at the same institution or researchers from institutions abroad. The problem with this arrangement was that fieldworkers from different institutions targeted the same matches, and spectators often became annoyed with fieldworkers outside the stadia. For example, the three largest universities in South Africa are situated in the City of Tshwane, and they all did fieldwork outside one stadium. Another example is a group of researchers who had a team of fieldworkers traveling to all ten stadia where games were hosted targeting all visiting nations. Unfortunately, this team of fieldworkers had to "compete" with fieldworkers from universities in all the host cities. In this sense, the extent to which research serves to forge connections between disciplines in a multidisciplinary research approach could be examined to (or "intending to") achieving a more collaborative spirit and better and more integrated research outputs.

References

Ademović, A., Čolakhodžić, E., Talović, M., & Kajmović, H. (2012). Top footballer model based on indicators of situational efficiency in round of 16 at the 2010 FIFA World Cup. *Homosporticus*, *14*(2), 33–40.

Ahmed, F., & Pretorius, L. (2010). Mega-events and environmental impacts: The 2010 FIFA World Cup in South Africa. *Alternation*, *17*(2), 274–296.

Alber, E., & Ungruhe, C. (2014). Fans and states at work: A Ghanaian fan trip to the FIFA World Cup 2010 in South Africa. *Soccer & Society*, *17*(1), 18–39.

Alegi, P. (2007). The political economy of mega-stadiums and the underdevelopment of grassroots football in South Africa. *Politikon*, *34*(3), 315–331.

Alegi, P. (2008). "A nation to be reckoned with": The politics of World Cup stadium construction in Cape Town and Durban, South Africa. *African Studies*, *67*(3), 397–422.

Allen, D. (2013). "The successes and challenges of hosting the 2010 FIFA World Cup": The case of Cape Town, South Africa. *Soccer & Society*, *14*(3), 404–415.

Allen, D., Knott, B., & Swart, K. (2013). "Africa's tournament"? The branding legacy of the 2010 FIFA World Cup™. *The International Journal of the History of Sport*, *30*(16), 1994–2006.

Almaraz, A., & Garcia, F. (2012). The lexicon sport of journalistic chronicles during the 2010 Football World Cup won by Spain in the Poem of the Cid. *Historia y comunicacion social*, *17*, 317–341.

Altheide, D. L. (1987). Reflections: Ethnographic content analysis. *Qualitative Sociology*, *10*(1), 65–77.

Angus, D., Rintel, S., & Wiles, J. (2013). Making sense of big text: A visual-first approach for analysing text data using Leximancer and Discursis. *International Journal of Social Research Methodology*, *16*(3), 261–267.

Aranke, S., & Zoller, K. (2010). Crisis management as representational strategy: The arrangement of "African" subjectivities and the 2010 World Cup. *Ecquid Novi: African Journalism Studies*, *31*(2), 131–148.

Baloyi, L., & Bekker, M. (2011). Causes of construction cost and time overruns: The 2010 FIFA World Cup stadia in South Africa. *Acta Structilia*, *18*(1), 51–67.

Bassa, Z., & Jaggernath, J. (2010). Living close to 2010 stadiums: Residents' perceptions of the 2010 FIFA World Cup and stadium development in Durban, South Africa. *Alternation*, *17*(2), 121–145.

Beres, L. K. (2011). Missed opportunities: HIV and the 2010 FIFA World Cup. *Public Health*, *125*(10), 725–726.

Berger, G. (2010). Image revisions: South Africa, Africa, and the 2010 World Cup. *Ecquid Novi: African Journalism Studies*, *31*(2), 174–190.

Berndt, A., Klopper, H. B., Niemann-Struweg, I., & Meintjes, C. (2013). Resident co-creation: The case of the 2010 Soccer World Cup. *European Business Review*, *25*(4), 336–350.

Bird, R., & Donaldson, R. (2009). "Sex, sun, soccer": Stakeholder-opinions on the sex industry in Cape Town in anticipation of the 2010 FIFA Soccer World Cup. *Urban Forum*, *20*(1), 33–46.

Bizzini, M., Schmied, C., Junge, A., & Dvorak, J. (2011). Precompetition medical assessment of referees and assistant referees selected for the 2010 FIFA World Cup. *British Journal of Sports Medicine*, *46*, 374–376.

Black, D. (2007). The symbolic politics of sport mega-events: 2010 in comparative perspective. *Politikon*, *34*(3), 261–276.

Blackshaw, T., & Coetzee, W. (2020). Re-imagining 'Event Management': On governmentality and its 'afterlife'. *International Journal of the Sociology of Leisure*, *3*, 369–387.

Bob, U., & Majola, M. (2011). Rural community perceptions of the 2010 FIFA World Cup: The Makhowe community in KwaZulu-Natal. *Development Southern Africa*, *28*(3), 387–399.

Bob, U., & Swart, K. (2009). Resident perceptions of the 2010 FIFA Soccer World Cup stadia development in Cape Town. *Urban Forum*, *20*(1), 47–59.

Bob, U., & Swart, K. (2010a). The 2010 FIFA World Cup and women's experiences in fan parks. *Agenda*, *24*(85), 85–96.

Bob, U., & Swart, K. (2010b). Sport events and social legacies. *Alternation*, *17*(2), 72–95.

Bohlmann, H. R., & Heerden, J. H. V. (2008). Predicting the economic impact of the 2010 FIFA World Cup on South Africa. *International Journal of Sport Management and Marketing*, *3*(4), 383–396.

Bolsmann, C. (2012). Representation in the first African World Cup: "World-class", Pan-Africanism, and exclusion. *Soccer & Society*, *13*(2), 156–172.

Bresler, N. C. (2011). Tourist considerations in hosting a mega sport event: 2010 FIFA World Cup in South Africa. *South African Journal for Research in Sport, Physical Education and Recreation*, *33*(2), 17–33.

Brook Napier, D. (2012). Globalization and transformation in South Africa: The hosting of the 2010 World Cup. *Education and Society*, *30*(1), 53–73.

Brown, L., Richards, S., & Jones, I. (2014). Sojourner perceptions of the St George Cross flag during the FIFA 2010 World Cup: A symbol of carnival or menace? *International Review for the Sociology of Sport*, *49*(1), 102–120.

Carreño Lara, E. (2012). Sport in the diplomatic arena: The case of FIFA World Cup South Africa 2010. *Estudios Políticos*, *41*, 170–188.

Casal, C. A., Maneiro, R., Ardá, T., Losada, J. L., & Rial, A. (2014). Effectiveness of indirect free kicks in elite soccer. *International Journal of Performance Analysis in Sport*, *14*(3), 744–760.

Cashman, R. (2005). *The bitter-sweet awakening. The legacy of the Sydney 2000 Olympic Games*. Walla Walla Press.

Chain, D., & Swart, K. (2010). Residents' perceptions of the 2010 FIFA World Cup: A case study of a suburb in Cape Town, South Africa. *Alternation*, *17*(2), 146–172.

Chappelet, J., & Junod, T. (2006). A tale of 3 Olympic cities: What can Turin learn from the Olympic legacy of other alpine cities? In D. Torres (Ed.), *Proceedings of Workshop on Major Sport Events as Opportunity for Development, 14–16 June, Valencia, Spain* (pp. 83–90).

Chistofaro, D. G. D., Orbolato, A., Orbolato, R., Picoli, R., & Fernandes, R. A. (2014). Aggressiveness in football: A quantitative analysis of the 2010 world cup matches. *Revista da Educacao Fisica*, *25*(3), 419–427.

Chuma, W. (2012). Framing the Cape Town World Cup stadium in the media: The politics of identity and sports in South Africa. *Journal of African Media Studies*, *4*(3), 315–329.

Collet, C. (2013). The possession game? A comparative analysis of ball retention and team success in European and international football, 2007–2010. *Journal of Sports Sciences*, *31*(2), 123–136.

Conradie, M. (2011). The 2010 FIFA World Cup as a commercial platform in print advertising. *African Identities*, *9*(3), 291–305.

Cornelissen, S. (2004). "It's Africa's turn!" The narratives and legitimations surrounding the Moroccan and South African bids for the 2006 and 2010 FIFA finals. *Third World Quarterly*, *25*(7), 1293–1309.

Cornelissen, S. (2007). Crafting legacies: The changing political economy of global sport and the 2010 FIFA World Cup™. *Politikon*, *34*(3), 241–259.

Cornelissen, S. (2008). Scripting the nation: Sport, mega-events, foreign policy and state-building in post-apartheid South Africa. *Sport in Society*, *11*(4), 481–493.

Cornelissen, S. (2011a). Mega event securitisation in a Third World setting glocal processes and ramifications during the 2010 FIFA World Cup. *Urban Studies*, *48*(15), 3221–3240.

Cornelissen, S. (2011b). More than a sporting chance? Appraising the sport for development legacy of the 2010 FIFA World Cup. *Third World Quarterly*, *32*(3), 503–529.

Cornelissen, S. (2012). "Our struggles are bigger than the World Cup": Civic activism, state-society relations and the socio-political legacies of the 2010 FIFA World Cup. *The British Journal of Sociology*, *63*(2), 328–348.

Cornelissen, S., Bob, U., & Swart, K. (2011). Towards redefining the concept of legacy in relation to sport mega-events: Insights from the 2010 FIFA World Cup. *Development Southern Africa*, *28*(3), 307–318.

Cornelissen, S., & Maennig, W. (2010). On the political economy of "feel-good" effects at sport mega-events: Experiences from FIFA Germany 2006 and prospects for South Africa 2010. *Alternation*, *17*(2), 96–120.

Cornelissen, S., & Solberg, E. (2007). Sport mobility and circuits of power: The dynamics of football migration in Africa and the 2010 World Cup. *Politikon*, *34*(3), 295–314.

Cornelissen, S., & Swart, K. (2006). The 2010 Football World Cup as a political construct: The challenge of making good on an African promise. *The Sociological Review*, *54*, 108–123.

Cotta, C., Mora, A. M., Merelo, J. J., & Merelo-Molina, C. (2013). A network analysis of the 2010 FIFA world cup champion team play. *Journal of Systems Science and Complexity*, *26*(1), 21–42.

Cronjé, F., van Wyk, D., & Botha, D. (2010). The 2010 FIFA World Cup – Service delivery, "afrophobia" and brand imperialism: Through the eyes of Frantz Fanon. *Alternation*, *17*(2), 297–322.

Cubizolles, S. (2011). Marketing identity and place: The case of the Stellenbosch Kayamandi Economic Corridor before the 2010 World Cup in South Africa. *Journal of Sport & Tourism*, *16*(1), 33–53.

Cubizolles, S. (2012). Integrating a popular sport into the patrimony in a South African provincial town: The case of football in Stellenbosch. *African Studies*, *71*(1), 108–126.

Cunefare, K. A., Ruhala, R., Ruhala, L., & Ortkiese, T. (2011). Vuvuzelas and their impact. *The Journal of the Acoustical Society of America*, *130*(4), 2564.

Darkey, D., & Horn, A. (2009). Homing in(n) on the economic benefits of 2010 FIFA World Cup: Opportunities for and misgivings of bed-and-breakfast operators in Gauteng, South Africa. *Urban Forum*, *20*(1), 77–91.

Davenport, T. H., & Prusak, L. (1998). *Working knowledge: How organizations manage what they know*. Harvard Business Press.

de Almeida, B. S., Bolsmann, C., Júnior, W. M., & de Souza, J. (2015). Rationales, rhetoric and realities: FIFA's World Cup in South Africa 2010 and Brazil 2014. *International Review for the Sociology of Sport*, *50*(3), 265–282.

Death, C. (2011). "Greening" the 2010 FIFA World Cup: Environmental sustainability and the mega-event in South Africa. *Journal of Environmental Policy & Planning*, *13*(2), 99–117.

Delva, W., Richter, M., De Koker, P., Chersich, M., & Temmerman, M. (2011). Sex work during the 2010 FIFA World Cup: Results from a three-wave cross-sectional survey. *PLoS One*, *6*(12), e28363.

Donaldson, R., Cornelissen, S., Swart, K., & Bob, U. (2008). Expectations of urban residents in the Western Cape on the 2010 FIFA World Cup. *Africa Insight*, *38*(3), 35–48.

Donaldson, R., & Ferreira, S. (2007). Crime, perceptions and touristic decision making: Some empirical evidence and prospects for the 2010 World Cup. *Politikon*, *34*(3), 353–371.

Dubin, S. C. (2011). Imperfect pitch. *African Arts*, *44*(2), 18.

Duk, O. S., Min, K. S., Kawczynski, A., Chmura, P., Mroczek, D., & Chmura, J. (2011). Endurance and speed capacity of the Korea republic football national team during the World Cup of 2010. *Journal of Human Kinetics*, *30*, 115–121.

du Plessis, S., & Maennig, W. (2011). The 2010 FIFA World Cup high-frequency data economics: Effects on international tourism and awareness for South Africa. *Development Southern Africa*, *28*(3), 349–365.

Duvall, S.-S., & Guschwan, M. C. (2013). Commodifying global activism and racial unity during the 2010 FIFA World Cup. *Communication, Culture & Critique*, *6*(2), 298–317.

Dvorak, J., Junge, A., Derman, W., & Schwellnus, M. (2011). Injuries and illnesses of football players during the 2010 FIFA World Cup. *British Journal of Sports Medicine*, *45*, 626–630.

e Castro, A. S. (2013). South Africa's engagement in sports diplomacy: The successful hosting of the 2010 FIFA World Cup. *The Hague Journal of Diplomacy*, *8*(3–4), 197–210.

Eirale, C., Farooq, A., Bisciotti, G., & Chalabi, H. (2013). How injuries affect rehabilitation workload in a National Team during an elite soccer event? An experience from the FIFA 2010 World Cup. *The Journal of Sports Medicine and Physical Fitness*, *53*(2), 192–197.

Eisenhauer, S., Adair, D., & Taylor, T. (2013). Fifa-isation: Security, brand protection and media management at the 2010 World Cup in South Africa. *Surveillance & Society*, *11*(4), 377–391.

Emser, M., & Francis, S. (2014). Media waves and moral panicking: The case of the FIFA World Cup 2010. *Strategic Review for Southern Africa*, *36*(2), 169.

Ferreira, S. (2011). South African tourism road to economic recovery: 2010 FIFA Soccer World Cup as vehicle. *Tourism Review International*, *15*(1), 91–106.

Ferreira, S. L., & Boshoff, A. (2014). Post-2010 FIFA Soccer World Cup: Oversupply and location of luxury hotel rooms in Cape Town. *Current Issues in Tourism*, *17*(2), 180–198.

Fieggen, A. G. (2010). World Cup 2010: A South African reflection. *World Neurosurgery*, *74*(2–3), 229–230.

Fonio, C., & Pisapia, G. (2014). Security, surveillance and geographical patterns at the 2010 FIFA World Cup in Johannesburg. *The Geographical Journal*, *18*(3), 242–248.

Formica, S. (1998). The development of festivals and special events studies. *Festival Management and Event Tourism*, *5*(3), 131–137.

Furin, J. J. (2011). Blood exposure at the 2010 International Federation of Football Association World Cup: Time for universal adaptation of universal precautions. *Clinical Journal of Sport Medicine*, *21*(3), 264–265.

Gau, L. S. (2013). Identifying core, exciting, and hybrid attributes in fans' assessments of major (world cup) spectator sports events. *Psychological Reports*, *113*(3), 786–803.

George, R., & Swart, K. (2012). International tourists' perceptions of crime-risk and their future travel intentions during the 2010 FIFA World Cup™ in South Africa. *Journal of Sport & Tourism*, *17*(3), 201–223.

Getz, D., & Page, S. J. (2015). Progress and prospects for event tourism research. *Tourism Management*, *52*, 593–631.

Giampiccoli, A., Lee, S., & Nauright, J. (2015). Destination South Africa: Comparing global sports mega-events and recurring localised sports events in South Africa for tourism and economic development. *Current Issues in Tourism*, *18*(3), 229–248.

Gibson, H. J. (1998). Sport tourism: A critical analysis of research. *Sport Management Review*, *1*(1), 45–76.

Gibson, H. J., Walker, M., Thapa, B., Kaplanidou, K., Geldenhuys, S., & Coetzee, W. (2014). Psychic income and social capital among host nation residents: A pre-post analysis of the 2010 FIFA World Cup in South Africa. *Tourism Management*, *44*(0), 113–122.

Ginsberg, R. (2010). World Cup 2010: An (Un) African World Cup. *Ecquid Novi: African Journalism Studies*, *31*(2), 191–204.

Giulianotti, R., & Klauser, F. (2010). Security governance and sport mega-events: Toward an interdisciplinary research agenda. *Journal of Sport & Social Issues*, *34*(1), 49–61.

GTQ, G. Q. (2011). Bars' and restaurants' compliance with the Guatemalan smoke-free law during the 2010 Soccer World Cup: A missed opportunity. *Tobacco Control*, *1*.

Guillodo, Y., & Saraux, A. (2013). Soccer World Cup 2010: Evaluation of portable ultrasonography for diagnosing musculoskeletal injuries. *Journal de Traumatologie du Sport*, *30*(4), 240–243.

Guinto Balanzar, G. (2010). World Cup South Africa 2010 and neurosurgery. *World Neurosurgery*, *74*(2), 231–232.

Haferburg, C. (2011). South Africa under FIFA's reign: The World Cup's contribution to urban development. *Development Southern Africa*, *28*(3), 333–348.

Hall, C. M., & Page, S. J. (2009). Progress in tourism management: From the geography of tourism to geographies of tourism–A review. *Tourism Management*, *30*(1), 3–16.

Hammett, D. (2011). British media representations of South Africa and the 2010 FIFA World Cup. *South African Geographical Journal*, *93*(1), 63–74.

Hammond, N. (2011). The disharmonious honking of the vuvuzelas: Homogenization and difference in the production and promotion of the 2010 Soccer World Cup in South Africa. *Soccer & Society*, *12*(1), 46–55.

Harper, D. (2001). *Online etymological dictionary* (Retrieved August 1, 2006).

Harris, L. (2011). Mega-events and the developing world: A look at the legacy of the 2010 Soccer World Cup. *South African Journal of International Affairs*, *18*(3), 407–427.

Heere, B., Walker, M., Gibson, H., Thapa, B., Geldenhuys, S., & Coetzee, W. (2013). The power of sport to unite a nation: The social value of the 2010 FIFA World Cup in South Africa. *European Sport Management Quarterly*, *13*(4), 450–471.

Hermann, U. P., Du Plessis, L., Coetzee, W. J., & Geldenhuys, S. (2013). Local residents' perceptions of the 2010 FIFA World Cup. *South African Journal for Research in Sport, Physical Education and Recreation*, *35*(1), 25–37.

Hicks, J. (2010). A gendered response to the 2010 World Cup: Insights from the South African Commission for gender equality. *Agenda*, *24*(85), 105–115.

Hiller, H. H. (2006). Post-event outcomes and the post-modern turn: The Olympics and urban transformations. *European Sport Management Quarterly*, *6*(4), 317–332.

Holtzhausen, D., & Fullerton, J. (2015). The 2010 FIFA World Cup and South Africa: A study of longer-term effects and moderators of country reputation. *Journal of Marketing Communications*, *21*(3), 185–198.

Jain, K., Bearden, J. N., & Filipowicz, A. (2013). Depression and forecast accuracy: Evidence from the 2010 FIFA World Cup. *International Journal of Forecasting*, *29*(1), 69–79.

Jazaerly, T., & Giorgadze, T. (2010). 2010 World Cup connections: Paul the Octopus in a lymph node aspirate. *International Journal of Surgical Pathology*, *18*(6), 521.

Jethro, D. (2014). Vuvuzela magic. *African Diaspora*, *7*(2), 177–204.

Jones, M. V., Coffee, P., Sheffield, D., Yangüez, M., & Barker, J. B. (2012). Just a game? Changes in English and Spanish soccer fans' emotions in the 2010 World Cup. *Psychology of Sport and Exercise*, *13*(2), 162–169.

Kaplanidou, K., Karadakis, K., Gibson, H., Thapa, B., Walker, M., Geldenhuys, S., & Coetzee, W. (2013). Quality of life, event impacts, and mega-event support among South African residents before and after the 2010 FIFA World Cup. *Journal of Travel Research*, *52*(5), 631–645.

Kaplanski, G., & Levy, H. (2014). Sentiment, irrationality and market efficiency: The case of the 2010 FIFA World Cup. *Journal of Behavioral and Experimental Economics*, *49*(0), 35–43.

Kempe, M., Vogelbein, M., Memmert, D., & Nopp, S. (2014). Possession vs. direct play: Evaluating tactical behavior in elite soccer. *International Journal of Sports Science*, *4*(6A), 35–41.

Kersting, N. (2007). Sport and national identity: A comparison of the 2006 and 2010 FIFA World Cups™. *Politikon*, *34*(3), 277–293.

Kim, J., Boo, S., & Kim, Y. (2013). Patterns and trends in event tourism study topics over 30 years. *International Journal of Event and Festival Management*, *4*(1), 66–83.

Kirby, S., Francis, B., & O'Flaherty, R. (2013). Can the FIFA world cup football (soccer) tournament be associated with an increase in domestic abuse? *Journal of Research in Crime and Delinquency*, online.

Klassen, T. P., Jadad, A. R., & Moher, D. (1998). Guides for reading and interpreting systematic reviews: I. getting started. *Archives of Pediatrics & Adolescent Medicine*, *152*(7), 700–704.

Knott, B., Allen, D., & Swart, K. (2012). Stakeholder reflections of the tourism and nation-branding legacy of the 2010 FIFA World Cup for South Africa. *African Journal for Physical Health Education, Recreation and Dance: Supplement 2*, *18*, 112–122.

Knott, B., Fyall, A., & Jones, I. (2015). The nation branding opportunities provided by a sport mega-event: South Africa and the 2010 FIFA World Cup. *Journal of Destination Marketing & Management*, *4*(1), 46–56.

Koortzen, P., & Oosthuizen, R. M. (2012). Psychological experiences in South African society before the 2010 FIFA World Cup from the systems psychodynamic and positive psychology perspectives. *SA Journal of Industrial Psychology*, *38*(2).

Krippendorff, K. (2012). *Content analysis: An introduction to its methodology*. Sage.

Kruger, E. A., & Heath, E. T. (2013). Along came a mega-event: Prospects of competitiveness for a 2010 FIFA World Cup™ host city. *Current Issues in Tourism*, *16*(6), 570–590.

Labuschagne, P. (2008). The impact of sport on nation building: A critical analysis of South Africa and the 2010 FIFA World Cup. *Africa Insight*, *38*(3), 3–14.

Lames, M., & Kolbinger, O. (2011). German Bundesliga Clubs' rankings in season 2010/11 were significantly affected by the number of players released to World Cup 2010. *International Journal of Performance Analysis in Sport*, *11*(2), 308–313.

Larlham, D. (2012). On empathy, optimism, and beautiful play at the First African World Cup. *TDR/The Drama Review*, *56*(1), 18–47.

Leggat, P. A., Shaw, M. T. M., & Toovey, S. (2010). Travelling to South Africa for the 2010 FIFA World Cup. *Travel Medicine and Infectious Disease*, *8*(2), 74–78.

Lepp, A., & Gibson, H. (2011). Tourism and world cup football amidst perceptions of risk: The case of South Africa. *Scandinavian Journal of Hospitality and Tourism*, *11*(3), 286–305.

Lopes, F., Loureiro, L. M., & Vieira, P. (2012). World Cup television. *Critical Arts*, *26*(5), 707–727.

Lopez, S. T. (2012). Race, place and soccer: Egypt, Morocco and "African" identity in the competition to host the 2010 FIFA World Cup. *Soccer & Society*, *13*(5–6), 639–652.

Machado, J. C., Barreira, D., & Garganta, J. (2013). Attacking efficacy and game pattern variability in soccer. *Revista Brasileira de Educação Física e Esporte*, *27*(4), 667–677.

Madichie, N. O. (2011). Whither the legacy of FIFA 2010 World Cup for entrepreneurship in South Africa? *African Journal of Business and Economic Research: The Implications of the South Africa 2010 FIFA World Cup for African Business: Special Issue 1*, *6*, 44–61.

Maennig, W., & Du Plessis, S. (2007). World Cup 2010: South African economic perspectives and policy challenges informed by the experience of Germany 2006. *Contemporary Economic Policy*, *25*(4), 578–590.

Maennig, W., & du Plessis, S. (2009). Sport stadia, sporting events and urban development: International experience and the ambitions of Durban. *Urban Forum*, *20*(1), 61–76.

Maennig, W., & Zimbalist, A. S. (Eds.). (2012). *International handbook on the economics of mega sporting events*. Edward Elgar Publishing.

Maharaj, A. (2010). The 2010 FIFA World Cup, sport events and tourism in Durban: Prospects and challenges. *Alternation*, *17*(2), 249–273.

Malerczyk, C., Nel, L., Gniel, D., & Blumberg, L. (2010). Rabies in South Africa and the FIFA Soccer World Cup: Travelers' awareness for an endemic but neglected disease. *Human Vaccines*, *6*(5), 385–389.

Manzo, K. (2012). Visualising modernity: Development hopes and the 2010 FIFA World Cup. *Soccer & Society*, *13*(2), 173–187.

Masemola, M. K., & Chaka, P. M. (2011). (T) races of terrorism beyond ports of entry: A retrospective assessment of the limits of profiling in the regulation of airport passenger traffic during the 2010 FIFA World Cup. *Social Dynamics*, *37*(1), 178–192.

Masuku, M. B. (2014). Liability to gambling fans and supporters in favour of fair play in soccer. *International Journal of Private Law*, *7*(1), 69–81.

Matheson, V. A., & Baade, R. A. (2004). Mega-sporting events in developing nations: Playing the way to prosperity? *South African Journal of Economics*, *72*(5), 1085–1096.

McKenna, F., & Bob, U. (2010). Business perceptions of the 2010 FIFA World Cup and related infrastructural development: A case study of the Moses Mabhida Stadium and the Durban beachfront development. *Alternation*, *17*(2), 200–223.

Mendelson, M., Davis, X. M., Jensenius, M., Keystone, J. S., von Sonnenburg, F., Hale, D. C., . . . GeoSentinel Surveillance, N. (2010). Health risks in travelers to South Africa: The GeoSentinel experience and implications for the 2010 FIFA World Cup. *The American Society of Tropical Medicine and Hygiene*, *82*(6), 991–995.

Menéndez, H., Bello-Orgaz, G., & Camacho, D. (2013). Extracting behavioural models from 2010 FIFA World Cup. *Journal of Systems Science and Complexity*, *26*(1), 43–61.

Molloy, E., & Chetty, T. (2015). The rocky road to legacy: Lessons from the 2010 FIFA World Cup South Africa stadium program. *Project Management Journal*, *46*(3), 88–107.

Muxel, S., Radmacher, M. C., Schäfer, S., Damaske, A., Münzel, T., & Gori, T. (2011). Changes in endothelial function in healthy volunteers during soccer World Cup 2010 games. *International Journal of Cardiology*, *152*(3), 395–396.

Nadvi, L. (2008). The ugly side of the beautiful game: The socio-economic impact of the 2010 FIFA World Cup on the city of eThekwini and its "poors". *World Journal of Managing Events*, *2*(1), 39–47.

Nassis, G. P. (2013). Effect of altitude on football performance: Analysis of the 2010 FIFA World Cup data. *The Journal of Strength & Conditioning Research*, *27*(3), 703–707.

Ndlovu-Gatsheni, S. J. (2011). Pan-Africanism and the 2010 FIFA World Cup in South Africa. *Development Southern Africa*, *28*(3), 401–413.

Newton, C. (2009). The reverse side of the medal: About the 2010 FIFA World Cup and the beautification of the N2 in Cape Town. *Urban Forum*, *20*(1), 93–108.

Ngonyama, P. (2010). The 2010 FIFA World Cup: Critical voices from below. *Soccer & Society*, *11*(1–2), 168–180.

Nicholson, M., Sherry, E., & Osborne, A. (2014). Negotiating national identity through loss: Australian newspaper coverage of the 2010 FIFA World Cup. *Soccer & Society* (ahead-of-print), 1–15.

Nicolau, J. L. (2012). The effect of winning the 2010 FIFA World Cup on the tourism market value: The Spanish case. *Omega*, *40*(5), 503–510.

Oh, M. (2013). "Hope for the win and hope for the defeat": Constructions of South Korean identity and the 2010 FIFA World Cup. *Soccer & Society*, *14*(5), 670–683.

Oosthuizen, R., & Koortzen, P. (2012). Psychological experiences in South African society before the 2010 FIFA World Cup. *International Journal of Psychology*, *47*, 634.

Pearce, P. L., & Wu, M. Y. (2016). Tourists' evaluation of a romantic themed attraction: Expressive and instrumental issues. *Journal of Travel Research*, *55*(2), 220–232.

Peeters, T., Matheson, V., & Szymanski, S. (2014). Tourism and the 2010 World Cup: Lessons for developing countries. *Journal of African Economies*, *23*(2), 290–320.

Pillay, U., & Bass, O. (2008). Mega-events as a response to poverty reduction: The 2010 FIFA World Cup and its urban development implications. *Urban Forum*, *19*(3), 329–346.

Pillay, V., & Salo, E. (2010). 2010 FIFA World Cup: Gender, politics and sport. *Agenda*, *24*(85), 4–10.

Polanyi, M. (1967). The tacit dimension New York. *Garden City*, 4.

Preuss, H. (2007). The conceptualisation and measurement of mega sport event legacies. *Journal of Sport & Tourism*, *12*(3–4), 207–228.

Quigg, Z., Hughes, K., & Bellis, M. A. (2013). Effects of the 2010 World Cup football tournament on emergency department assault attendances in England. *European Journal for Public Health*, *23*(3), 383–385.

Ramma, L. (2011). Vuvuzela media coverage during the 2010 FIFA soccer world cup tournament: Impact on raising awareness of noise-induced hearing loss. *Noise and Health*, *13*(55), 415.

Reppel, M., Franzen, K., Bode, F., Weil, J., Kurowski, V., Schneider, S. A., . . . Mortensen, K. (2013). Central hemodynamics and arterial stiffness during the finals of the world cup soccer championship 2010. *International Journal of Cardiology*, *166*(3), 627–632.

Richter, M. L., Chersich, M. F., Scorgie, F., Luchters, S., Temmerman, M., & Steen, R. (2010). Sex work and the 2010 FIFA World Cup: Time for public health imperatives to prevail. *Global Health*, *6*, 1.

Richter, M. L., Luchters, S., Ndlovu, D., Temmerman, M., & Chersich, M. F. (2012). Female sex work and international sport events-no major changes in demand or supply of paid sex during the 2010 Soccer World Cup: A cross-sectional study. *BMC Public Health*, *12*(1), 763.

Richter, M. L., Scorgie, F., Chersich, M. F., & Luchters, S. (2014). There are a lot of new people in town: But they are here for soccer, not for business' a qualitative inquiry into the impact of the 2010 soccer world cup on sex work in South Africa. *Global Health*, *10*, 45.

Roberts, D. (2010). Durban's future? Rebranding through the production/policing of event-specific spaces at the 2010 World Cup. *Sport in Society*, *13*(10), 1486–1497.

Roelofse, C. (2011). Human trafficking as an organised crime in South Africa: Pre-, during and post Soccer World Cup 2010. *Acta Criminologica*, *24*(3), 1–19.

Rowe, D., & Baker, S. A. (2012). The "fall" of what? FIFA's public viewing areas and their contribution to the quality of public life. *Space and Culture*, *15*(4), 395–407.

Ryynänen, J., Dvorak, J., Peterson, L., Kautiainen, H., Karlsson, J., Junge, A., & Börjesson, M. (2013). Increased risk of injury following red and yellow cards, injuries and goals in FIFA World Cups. *British Journal of Sports Medicine*, *47*(15), 970–973.

Ryynänen, J., Junge, A., Dvorak, J., Peterson, L., Karlsson, J., & Börjesson, M. (2013). The effect of changes in the score on injury incidence during three FIFA World Cups. *British Journal of Sports Medicine*, *47*(15), 960–964.

Ryynänen, J., Junge, A., Dvorak, J., Peterson, L., Kautiainen, H., Karlsson, J., & Börjesson, M. (2013). Foul play is associated with injury incidence: An epidemiological study of three FIFA World Cups (2002–2010). *British Journal of Sports Medicine*, *47*(15), 986–991.

Saayman, M., & Rossouw, R. (2008). The economic value of the 2010 Soccer World Cup. *Acta Commercii*, *8*(1), 1–14.

Saurombe, A., & Saurome, N. M. (2011). Taking stock of the 2010 FIFA World Cup: Did it live up to expectations of being "the African World Cup"? *Commonwealth Youth and Development*, *9*(2), 50–59.

Sinclair, I., & Matlala, M. (2011). The use of technology and leadership in enhancing strategic cooperative policing within the SADC region. *International Journal of African Renaissance Studies-Multi-, Inter- and Transdisciplinarity*, *6*(1), 47–59.

Smit, J. A. (2010). Dennis Brutus: Activist for non-racialism and freedom of the human spirit. *Alternation*, *17*(2), 8–71.

Smith, A. (2005). Reimaging the city: The value of sport initiatives. *Annals of Tourism Research*, *32*(1), 217–236.

Solari, C. A., & Solomons, L. (2012). The World Cup effect: Using football to engage men with dementia. *Dementia*, 1471301211429177.

Stone, C., & Rod, M. (2014). Unfair play in World Cup qualification? An analysis of the 1998–2010 FIFA World Cup performances and the bias in the allocation of tournament berths. *Soccer & Society* (ahead-of-print), 1–18.

Swart, K., & Bob, U. (2007). The eluding link: Toward developing a national sport tourism strategy in South Africa beyond 2010. *Politikon*, *34*(3), 373–391.

Swart, K., Bob, U., & Turco, D. (2010). Media, crime and the 2010 Soccer World Cup in South Africa: Pre-event analysis and perceptions. *Alternation*, *17*(2), 225–248.

Swart, K., & Bob, U. (2012). The methodology of mega-events: The culture of vanity and measuring real benefit. In D. Hassan & J. Lusted (Eds.), *Managing sport social and cultural perspective* (pp. 252–267). Routledge.

Swart, K., Linley, M., & Bob, U. (2013). The media impact of South Africa's historical hosting of Africa's first mega-event: Sport and leisure consumption patterns. *The International Journal of the History of Sport*, *30*(16), 1976–1993.

Tichaawa, T. M., & Bob, U. (2015). Leveraging mega-events beyond the host nation: A case study of the 2010 FIFA World Cup African Legacy Programme in Cameroon and Nigeria. *Leisure Studies*, 1–16.

Tichaawa, T. M., & Swart, K. (2009). Cameroonian fans' perceptions of the 2010 FIFA World Cup: A case study of Buea and Limbe. *Alternation*, *17*(2), 173–199.

Tomlinson, R., Bass, O., & Bassett, T. (2011). Before and after the vuvuzela: Identity, image and mega-events in South Africa, China and Brazil. *South African Geographical Journal*, *93*(1), 38–48.

Tscholl, P. M., & Dvorak, J. (2012). Abuse of medication during international football competition in 2010 – lesson not learned. *British Journal of Sports Medicine*, *46*(16), 1140–1141.

Van Blerk, L. (2011). "Managing" Cape Town's street children/youth: The impact of the 2010 World Cup bid on street life in the city of Cape Town. *South African Geographical Journal*, *93*(1), 29–37.

van der Meij, L., Almela, M., Hidalgo, V., Villada, C., Ijzerman, H., van Lange, P. A. M., & Salvador, A. (2012). Testosterone and cortisol release among Spanish soccer fans watching the 2010 World Cup final. *PLoS One*, *7*(4), e34814.

Van Der Westhuizen, J. (2007). Glitz, glamour and the Gautrain: Mega-projects as political symbols. *Politikon*, *34*(3), 333–351.

van der Westhuizen, J., & Swart, K. (2011). Bread or circuses? The 2010 World Cup and South Africa's quest for marketing power. *The International Journal of the History of Sport*, *28*(1), 168–180.

van der Zee, E., & Go, F. M. (2013). Analysing beyond the environmental bubble dichotomy: How the 2010 World Cup case helped to bridge the host-guest gap. *Journal of Sport & Tourism*, *18*(3), 161–183.

van Wyk, J.-A. (2014). The 2010 FIFA World Cup in South Africa: The politics of hosting a mega international event. *World Journal of Managing Events*, *2*(1), 1–9.

Venter, R. B., Rogerson, C. M., Semens, A., & Myres, K. (2012). FIFA World Cup 2010: Implications for and effects on the entrepreneurial performance of South Africa's informal sector. *African Journal of Business Management*, *6*(1), 449–458.

Vincent, J., & Hill, J. (2011). Flying the flag for the En-ger-land: The Sun's (re) construction of English identity during the 2010 World Cup. *Journal of Sport & Tourism*, *16*(3), 187–209.

Vincent, J., Kian, E. M., Pedersen, P. M., Kuntz, A., & Hill, J. S. (2010). England expects: English newspapers' narratives about the English football team in the 2006 World Cup. *International Review for the Sociology of Sport*, *45*(2), 199–223.

Waardenburg, M., van den Bergh, M., & van Eekeren, F. (2015). Local meanings of a sport mega-event's legacies: Stories from a South African urban neighbourhood. *South African Review of Sociology*, *46*(1), 87–105.

Walker, M., Kaplanidou, K., Gibson, H., Thapa, B., Geldenhuys, S., & Coetzee, W. (2013). "Win in Africa, with Africa": Social responsibility, event image, and destination benefits. The case of the 2010 FIFA World Cup in South Africa. *Tourism Management*, *34*(0), 80–90.

Wallace, J. L., & Norton, K. I. (2014). Evolution of World Cup soccer final games 1966–2010: Game structure, speed and play patterns. *Journal of Science and Medicine in Sport*, *17*(2), 223–228.

Weed, M. (2006). Sports tourism research 2000–2004: A systematic review of knowledge and a meta-evaluation of methods. *Journal of Sport & Tourism*, *11*(1), 5–30.

Yancey Ii, A. H., Fuhri, P. D., Pillay, Y., & Greenwald, I. (2008). World Cup 2010 planning: An integration of public health and medical systems. *Public Health*, *122*(10), 1020–1029.

Zroback, C., Levin, D., Manlhiot, C., Alexander, A., van As, A. B., & Azzie, G. (2014). Impact of the 2010 FIFA (Federation Internationale de Football Association) World Cup on pediatric injury and mortality in Cape Town, South Africa. *The Journal of Pediatrics*, *164*(2), 327–331.

Zuckerman, J. N., Bröker, M., & Worth, C. (2010). 2010 FIFA World Cup South Africa: Travel health issues and new options for protection against meningococcal disease. *Travel Medicine and Infectious Disease*, *8*(2), 68–73.

12

MANAGEMENT OF CULTURAL ROUTES AS THE NEW STATUS QUO IN URBAN CULTURAL TOURISM

Theodor Rachiotis and Panoraia Poulaki

Introduction

Over the past decade, tourism has benefited from the allocation of subsidies through European programs, while structural funds and other EU programs have supported the development of businesses, services, and infrastructure in the tourism sector (Commission of the European Communities, 2006). These packages concerned sustainable tourism and support for sustainable forms of tourism, such as cultural tourism. Cultural tourism is directly linked to the preservation and upgrading of cultural and natural heritage, involving more and more sectors and shaping new forms of tourism (Št´Astná et al., 2020). The European Regional Development Fund is responsible for implementing the financial instruments in this tourism sector, achieving the protection of the cultural and natural heritage, the upgrading of tourism infrastructure, the promotion of ICT, as well as common cross-border tourism strategies between regions (The European Parliament, 2021).

One of the emerging forms of tourism in recent years and directly related to cultural tourism is also urban tourism, which arose from the recognition of the rallying of visitors to urban centers and the improvement of the benefits offered in city-break destinations. Competitive holiday packages meet the expanded tourist needs, while providing for the sustainable development of the areas and monuments (D'Auria, 2009; Righini, 2020).

A characteristic form of connection between urban and cultural tourism is the cultural routes, which concern tours, which are related to the environment, acquaintance with the local tradition and the visit to historical monuments and settlements. At the same time, as a form of alternative tourism, it includes the participation of tourists in cultural events. Cultural routes give special value cultural and economic values to the regions, while contributing to their sustainability,

DOI:10.4324/9781032633374-14

forming a brand. The goals of the routes are to attract tourists, lengthen the tourist season, and create growth conditions for businesses and producers operating in each place (Banila, 2016).

Policies to support urban cultural tourism in Europe

It is not surprising that the "green city" is considered an asset for the tourism development of European cities (Angelidou et al., 2017). The European Green Capital Award promoted by the European Commission acts as a branding platform that delivers advantages in terms of increasing tourism (ec.europa.eu, 2020). From the point of view of resilient cities, the challenge lies in adapting to climate change, since higher temperatures and extreme weather events demonstrate the vulnerability of urban systems in coastal zones (Kakoutis, 2019). Also, the perspective of "Creative Cities" intersects with city tourism, especially in the case of the European Capital of Culture (ECoC) system based on culture as a trigger for smart, sustainable and inclusive growth (OECD, 2014). The program has recently been strengthened with its more stable integration into the overall urban cultural strategy, to broaden and expand the local heritage of the European Capital of Culture. In this context, sustainable tourism is considered as the natural result of the ECoC program (Triarchi & Karamanis, 2017).

The contribution of a Destination Marketing and Management Organization (DMMO) is key to ensuring the sustainability of a destination. DMMO guides, organizes, and manages the parties involved and processes to achieve the objectives of the marketing and promotion plan. Its target is to create a competitive advantage for packages and the destination as a whole (Eichelberger et al., 2020). Specific benefits of such a structure in the organization and development of urban cultural tourism are the creation of strong collaborative networks, which can lead to short-term and long-term advantages for destinations, the increase in tourist demand through the provision of incentives, the widening of seasonality, the activation of the local mechanism with the involvement of enterprises, with the corresponding benefits of generating income, and strengthening the local economy and employability. Also, advantages of the existence of a DMMO are the promotion of the cultural and architectural wealth of the regions, the partnership between the private and public sectors, forming the ground for easier undertaking of similar projects in future and a framework for cooperation and efficient interaction, as well as the effective coordination of the actors involved, with greater chances of achieving the objectives (Clarke, 1997).

More specifically, regarding PPPs in the field of destination promotion, the advantages are very significant, since it is possible to transfer know-how, specialization, and flexibility from the private sector to the public sector. PPPs also help the private sector to boost employability, leading to the growth of the economy in general. In addition, PPPs are free from time-consuming bureaucratic procedures (Koumelis, 2006). PPPs achieve faster construction and management of tourism infrastructure, achieving a higher level of visitor satisfaction and increasing revenues and quality of life for the local community. Finally, the use of public goods also ensures the sustainability of projects, industries, and sectors involved (INSETE, 2020).

In addition to the private and public bodies that will be involved, the local community, businesses and investors, the role of municipalities is crucial. The Municipalities are the ones that will mediate between the bodies and the formulation of the strategy (Ministry of Tourism, 2021). They cooperate with the DMMO by offering it the means, tools, staff, and directions to support its responsibilities. Very important is the role of Municipalities in raising awareness of permanent residents and in supporting complementary actions for the promotion of urban tourism and the promotion of cultural identity (E.E.T.A.A., 2020). However, overall success is judged on the effectiveness of communication, which will allow municipalities and DMMO the smooth and continuous guidance and support of residents and local businesses.

The concept of cultural route as a means of highlighting and managing cultural heritage

One element, which is directly linked to cultural heritage, is the possibility of its sustainable development. Sustainability, moreover, is not limited only to material heritage, but also to intangible cultural and natural, that is, to the protection of the environment. The integrated management of cultural heritage is today perhaps the most important strategic choice for local regions; since it is considered the core of the cultural identity and history of the people, it functions as a unifying element for people, but also as a ground for the development of creativity, as well as an object of social responsibility (Charalabopoulou, 2019).

Cultural routes play a decisive role in the effective and sustainable management of cultural heritage, which enable not only monuments and archaeological sites to be exploited with economic benefit, but also to create a protective framework for them. Cultural routes aim at the reorganization, decentralization, dispersion and redistribution of tourist traffic, and the promotion of a more uniform, balanced, and sustainable local and regional economic development (MCA, 2021). The design of the routes, however, requires the adaptation of plans and projects according to the needs of the place and its peculiarities. The resulting products will also be a sample of the cultural reserve and historical wealth that a region has, while for the success of the project, it is necessary to link the cultural heritage with tourism and attract visitors (Ashworth & Page, 2011; Lagos, 2012).

The cultural route goes beyond the narrow spatiotemporal context and takes on diachronic dimensions, incorporating the cultural identity and history of the place. Through the creation of routes, it is possible to shape economically and socially sustainable experiences, using cultural heritage spatially and urban planning based on sustainability. At the same time, as a tourism product, it is required to be competitive and offer personalized experiences (Pappa, 2018). In the relationship between the management of cultural heritage and the development of cultural routes, it could be noted that the latter increases their competitiveness, as tourist products, as long as the impact they have on the environment is minimized. Other factors that consent to this are the increase in employability, the support of the local economy, and the containment of the rural population (Somarakis et al., 2016). For this reason, there are many cases where in the design of cultural routes and experiences the involvement of locals, business owners, and social groups is encouraged, in order to enrich the products with authentic narratives and characteristics (Briedenhann & Wickens, 2004).

The design and promotion of cultural routes is a collective initiative, with the necessary cooperation of institutions, citizens, and social organizations. In addition, businesses can be the "key" to the effective support of these tourism products (Fergadakis, 2021). The operation of businesses offering tourist routes to cultural monuments would be an interesting proposal, which requires, however, careful study of the existing forces and risks (Diazoma, 2022). An approach that has been on the rise in recent years is clusters, that is, clusters of businesses, which are a leading example of a policy to enhance entrepreneurship and hold a prominent position in the development agenda of the most developed economies. Clusters of enterprises are usually created in urban centers and function as landmarks and centers providing high-quality financial services and high-tech companies or centers of cultural activities and places of consumption. The advantages they offer are the promotion of routes and the attraction of visitors, as well as the support of communication and promotion of a place (Duarte-Duarte et al., 2021).

In the case of tourist routes to cultural monuments, taking on the task of promoting experiences by businesses, would have significant positive results. Initially, the responsibilities are shared since the companies have the exclusive responsibility of managing the cultural routes. Having the necessary tools, they can attract independent visitors or attract those who visited each monument. At the same time, they function as a "showcase" for the monument and the tourist product. Finally, the inputs resulting from sales strengthen the local economy, while cooperation with local businesses creates new jobs, improving the overall image of the region (Gosal et al., 2019; Nilashi et al., 2019).

A typical example is the "Within Athens" Program, during which "*personalized advisory support actions were created to enhance entrepreneurship and innovation*," to support small and medium-sized enterprises, in the context of the development of tourism/business routes (Diazoma, 2022). This program combined the promotion of the cultural identity and history of a place, combined with elements of entrepreneurship. This enabled the involvement of local small- and medium-sized enterprises and the creation of tourism products, related to routes (Guo et al., 2019).

Cultural routes as an integrated territorial investment project

The exploitation of cultural routes has been placed at the center of interest of states, in recent years, because it is a tool for spatial development, promotion of areas, support of the local and national economy, and promotion of cultural identity. At this point, the Integrated Territorial Investments meet, which concerns an instrument of "*integrated spatial development for the implementation of development strategies of specific spatial units*" (NSRF, 2015). It is based on innovative approaches aimed at promoting the sustainable development of areas, as well as the management of historical and cultural monuments (Banila, 2016).

More specifically, ITIs are a tool, funded by European funds and programs, thus overcoming bureaucratic rigidities. The content of such initiatives mainly concerns addressing problems identified in specific areas, making these places not just visitable, but poles of attraction for tourists. To make this possible, a spatial development plan for each region is being designed, which is developed in stages and includes defined objectives and results, including the maintenance of buildings, the improvement of the economic situation and employment, and the preservation of the natural and built environment based on a sustainable approach (Makrigiorgou, 2016).

It could be noted that this approach is of great interest and benefit to all who are involved, because experiential or thematic cultural routes are utilized in this way, which are a different tourist proposal, but also competitive packages. It is a holistic approach to highlighting cultural heritage and diversity, enhancing the attractiveness of the regions. At the same time, technological actions can be integrated and collaborating parties can be involved by creating a separate product. The role of ITI's strategic plans is of crucial importance since in this way there is publicity, greater involvement of institutions and businesses, transparency is ensured and the effectiveness of interventions related to sustainability is ensured (Region of Western Greece, 2016).

Cultural routes as a tourist product in Greece (funding from the NSRF and the recovery fund)

The purpose of an NSRF program in tourism is the qualitative upgrade of tourism products and services at all levels, to increase demand. The pandemic has had a significant impact on all sectors, especially on the tourism industry. The conditions that were formed led to the creation of new NSRF programs (2014–2020 and 2021–2027) to support businesses. In the post-Covid-19 era, emphasis is placed on maritime and spa tourism, ecotourism, as well as cultural tourism (TornosNews.gr, 2021). At the same time, thematic tourism products are promoted, while the strengthening of the establishment of tourist-visitable farms is provided for. It is understood that special tourism is no longer just an alternative option or trend, but a necessity for the sustainability of tourism businesses and the achievement of national targets in the sector (De Araujo & Bramwell, 2002).

Also, the Recovery Fund is a great development opportunity for tourism. Greece is expected to receive resources of 65 billion euros from the European Union, which will be allocated for the modernization of the Greek economy, its energy and digital transition, and the development of innovation (Malliara, 2021; Kaczorowska-Ireland, 2021). As far as alternative tourism is concerned, EUR 0.6 billion will be spent on strengthening its individual forms, as well as on training the staff employed. In addition to these initiatives, projects are expected aimed at highlighting new archaeological sites and other points of cultural interest, developing new roads, ports–marinas, and waterways. In this way, the infrastructure will be upgraded and the domestic tourism product will be supported, through the attraction of new groups of tourists and the fight against seasonality. The World Tourism Organization claims that by 2030, global tourism mobility will increase by 20%, with Greece claiming a high position in tourist preferences (OT.gr, 2021).

Cultural routes in whichever country they are implemented, have common features, such as a key theme, from which the individual actions are triggered, forming a brand. For this reason, the European Recovery Fund recently announced a program, in which Greece is expected to participate, receiving 650 million euros, to upgrade services and monuments of cultural interest, stimulating their visitors and promotion. For this program, five thematic cultural routes have already been designed, telling stories from the second millennium BC, and covering both historical and religious themes. The routes are as follows: 1) In the footsteps of the Apostle Paul, 2) The Cultural Egnatia, 3) The road to the West, with a subtitle "From Homer to Cervantes," 4) Monumental works of Cyclops and people during the Mycenaean era (second millennium BC), and 5) Network of castles: From Byzantium to the Ottoman domination (www.culture.gov.gr, 2022).

These routes are not limited only to the construction of stories and the provision of an interesting tour for those who will choose them but are accompanied by a series of interventions, the creation of infrastructure and the development of initiatives for the overall support, and promotion of the project. The goals are to attract tourists, lengthen the tourist season, and create development conditions for businesses and producers. At the same time, they aim to attract investors, in order to preserve monuments and archaeological sites, and to stimulate the local economy, through the securing of resources (Charami, 2021; Mitoula & Kaldis, 2020).

Digital cultural routes. Moving into a new era

The recognition of the importance of cultural routes at the local and national levels, as well as the awareness of the ever-changing socio-economic environment, have led to the creation of innovative tourism products, such as digital cultural routes (Góral, 2016). Now a variety of technological tools are available, allowing a visit to a monument or a museum from any part of the planet. The possibilities offered through ICT for creativity and experimentation are numerous; however, for a product to be considered successful, it is necessary to meet the needs and expectations of visitors. New technologies, however, make public access easier, even on cultural routes. For this reason, the sustainable digital transformation of cultural reserves in the new era should be considered, but with an emphasis on a balance between physical presence and virtual representation (Poulaki et al., 2021).

Thousands of visitors are attracted each year by the cultural routes, with some focusing on the religious aspects, mainly in relation to the pilgrimage, others on the special cultural and natural heritage of the individual cultural routes, analyzed from the point of view of the history of art or geography. However, today the availability of new data sources, especially in digital form, offers innovative ways of analyzing these cultural objects (Economou, 2019). On the one hand, there are new perspectives offered by open data sets about the areas crossed by the routes, which can be intersected in an interesting way with more conventional fieldwork. On the other hand, since walkers are increasingly making extensive use of their mobile phones to share their experiences on social media platforms or through ad hoc applications, analyzing their interactions online can provide a real-time picture of what is happening along these paths (Poulaki et al., 2021).

Through this amount of available data, cultural routes now have a digital presence. The institutions involved in them have created procedures for the digitization and exchange of information about these cultural objects: from national databases to archives from museum catalogs, lists of UNESCO, or other institutions on any territorial scale. All these digital documents now facilitate the study and preservation of cultural heritage. In addition, cultural heritage actors can rely on new media to share their cultural practices or create new ones by posting their photos on Instagram, sharing their experiences on a Facebook page, or enriching a Wikipedia page (Berry, 2012).

The common character of the new methods of visualization of cultural sites lies in the framing of information, by combining topographical elements. The topographic map can be defined as an "upgraded model representing the relief determined by the altitude and human settlements of a geographical area with precision and detail in a horizontal plane" (IGN). The visualization of digital information based on graphs, such as an internet map or a graph of the coexistence of words, gathers topological (relational) and topographical (contextual) elements in a digital spatial representation, which creates a new digital reality for urban cultural tourism (Connor, 2004).

Methodology

For this research, a quantitative approach was considered more appropriate, with data measurements both qualitative and quantitative, which are then analyzed using standardized statistical techniques, such as descriptive and inductive statistics. The main research tool was the structured questionnaire (Creswell, 2016). The selection of participants was made critically, and the sampling applied is characterized as intentional (Cohen et al., 2008). The research questions aim 1) To analyze the cultural routes as a factor of sustainable management of cultural wealth and development of urban cultural tourism. 2) To present the way in which Private-Public Partnerships, but also bodies such as the NSRF, the Recovery and Resilience Fund (Next Generation EU), or the DMMOs contribute to the strengthening and promotion of cultural routes.

The collection of the data started in October 2021 and was completed in December 2021. The survey focused on people associated with tourism development and promotion, entrepreneurship, and accommodation. The participants' responses were statistically analyzed in the Social Package for the Social Sciences (SPSS) 25.0.0.

The survey was based on 75 people, 56 women and 19 men; 68% of the participants were between 36 and 55 years old. At the same time, 56% belonged to married people and 44% to unmarried people. Regarding the level of education, 54.7% had a master's degree and 34.7% had a university degree/TEI. At the same time, 62.7% of the sample belonged to employees and 22.7% to freelancers. As for the institutions or businesses of work, 17.3% of the sample belonged to ministers – deputy ministers – general secretaries and to the administrative staff of tourism and culture, 13.3% belonged to people associated with the Municipality of Athens and 9.3% of the participants were related to catering businesses, 8% were tourism enterprises, and the same percentage were related to cultural enterprises.

Results

Sources of funding

Initially, the participants were asked about the extent to which the public private partnership (PPP) is promoted in Greece in matters of culture and tourism. The analysis showed that 52% of the sample answered rarely and 40% answered frequently. About the sources of funding for the enhancement and promotion of cultural routes, there was a possibility of multiple choice.

Of the 87 replies, 34.5% concerned community funds (n = 30) and 23% were related to national resources (n = 20). Also, 18.4% were related to a combination of sources (n = 16) and 11.5% were related to private resources (n = 10). At the same time, it is worth noting that 45.2% (n = 33) answered that in some cases, perhaps, the autonomy of local government organizations from the state for the easier finding of resources would have a positive effect on the Municipality and the Region in relation to cultural routes.

In addition, participants were asked to justify their response and of the 16 justifications, 56.3% were related to the fact that the autonomy of LGOs would be functional, but if there were the right people in the right places, the appropriate organization, the appropriate funds and the appropriate organization (n = 9). At the same time, six people noted that the autonomy of local authorities already exists and is considered important for finding resources (37.5%).

Sustainability and future directions

Initially, it is noted that 94.7% (n = 71) of the sample agreed that cultural routes can become useful tools to promote the sustainable development of areas. 29.3% (n = 22) agreed that actions have been taken to link the protection of the environment with cultural routes. At the same time, 52% answered that the integration of technologies in the cultural routes themselves is necessary and 32% answered that it is very important (n = 63). In addition, 50% (n = 9) noted that technologies enhance the tourist's experience, add attractiveness and attract a wider audience and 38.9% (n = 7) noted that they offer additional services such as digital maps, GPS, social network elements, QR codes and self-guided tours. Finally, 74.5% (n = 41) of the 55 participants who answered the relevant question stated that the impact of the "Green Cultural Routes," axed by citizens, tourists, institutions, and the overall socio-economic sign, is positive.

Inductive statistics

Then, cross tables were created for combinations of questions of the questionnaire. It is worth noting that 9 of the 20 people who demonstrated as a source of funding for the strengthening and promotion of cultural routes the national resources, 13 out of the 29 people who pointed to community resources as a

TABLE 12.1 Pivot table of descriptive statistics

Can cultural routes become useful tools to promote the sustainable development of areas?	I agree (94,7%)
Have specific actions been taken to link environmental protection to cultural routes?	Yes (29,3%) – NA/NA (53,3%)
How important is the integration of technologies in the cultural routes themselves (digitization, augmented reality) considered?	Necessary (52%) – Very important (32%)
If you wish, please specify your answer.	They enhance the tourist's experience, add attractiveness and attract a wider audience (50%) – They offer additional services such as digital maps, GPS, social network elements, QR codes and self-guided tours (38.9%)
How would you characterize the impact of the "Green Cultural Routes," with the axes of citizens, tourists, institutions and the overall socio-economic sign?	Positive impact (74,5%)

TABLE 12.2 Sources of funding for the strengthening and promotion of cultural routes and views on the autonomy of Local Government Organizations from the state

		In your opinion, would the autonomy of local government organizations from the state to make it easier for the European Union to find resources from the European Union have any positive effect on the Municipality and the Region in relation to cultural routes?					
		No	*Yes*	*In some cases, perhaps*		*NA/NA*	*Total*
What sources of funding are chosen to support and promote cultural routes?	National resources		3	9	7	1	20
	Community resources		4	13	11	1	29
	Private resources		1	3	6	0	10
	Combining sources		3	6	5	2	16
	NA/NA		1	3	6	1	11
Total		6	20	23		3	52

source of funding and 6 out of the 16 people who pointed to the combination of sources as a source of funding replied that the autonomy of Local Government Organizations would have a positive effect on cultural routes. On the other hand, 6 of the 10 people who pointed to private resources as a source of funding and 6 out of 11 people who did not demonstrate a source of funding replied that the autonomy of local authorities would have a positive effect, but in some cases.

It was also found that 22 of the 71 who agree that cultural routes can become useful tools to promote the sustainable development of areas replied that specific actions have been taken to link environmental protection with cultural routes. On the contrary, 13 of the 71 who again agree that cultural routes can become useful tools for promoting the sustainable development of areas replied that no specific actions have been taken to link the protection of the environment with cultural routes.

In the results, it was observed that 60 of the 71 people who stated that cultural routes can become useful tools to promote the sustainable development of areas, responded that the integration of technologies is very important or necessary. Therefore, there is a strong tendency in the sample to support technologies on cultural routes

It is noted that of the 54 people who agreed that cultural routes can become useful tools to promote the sustainable development of areas, 41 demonstrated a positive impact of cultural routes. Moreover, of these 54 people, only four showed a moderate impact. Therefore, the sample showed a strong attitude of positive impact

TABLE 12.3 Cultural routes as useful tools for promoting the sustainable development of areas in relation to actions to link environmental protection with cultural routes

		Have specific actions been taken to link environmental protection to cultural routes?			
		No	*Yes*	*NA/NA*	*Total*
Cultural routes can become useful tools to promote the sustainable development of areas. What is your opinion on this?	I don't agree	0	0	1	1
	I agree	13	22	36	71
	NA/NA	0	0	3	3
Total		13	22	40	75

TABLE 12.4 Cultural routes as useful tools to promote the sustainable development of areas in relation to the importance of integrating technologies into the cultural routes themselves.

		How important do you consider the integration of technologies in the cultural routes themselves (digitization, augmented reality)?				
		Not important	*Important*	*Very important*	*Necessary*	*Total*
Cultural routes can become useful tools to promote the sustainable development of areas. What is your opinion on this?	I don't agree	0	0	0	1	1
	I agree	2	9	23	37	71
	NA/NA	0	1	1	1	3
Total		2	10	24	39	75

of green cultural routes, while at the same time showing a positive attitude toward whether cultural routes can become useful tools to promote the sustainable development of areas.

The results then showed a statistically important relationship between the questions about the role of Destination Marketing and Management Organizations in promoting cultural routes and the importance of integrating technologies into cultural routes ($r = .244$, $p = .035$)

TABLE 12.5 Correlations of the role of Marketing and Destination Management Organizations, the promotion of PPPs and the importance of integrating technologies.

		How important do you consider the integration of technologies in the cultural routes themselves (digitization, augmented reality)?
What is the role of Destination Marketing and Management Organizations in the promotion of cultural routes?	Pearson Correlation	,244*
	Sig. (2-tailed)	,035
	N	75
The PPP (public–private partnership) is promoted in matters of culture and tourism, as shown by European programs and independent initiatives. To what extent is this direction chosen in Greece?	Pearson Correlation	,198
	Sig. (2-tailed)	,089
	N	75
How important do you consider the integration of technologies in the cultural routes themselves (digitization, augmented reality)?	Pearson Correlation	1
	Sig. (2-tailed)	
	N	75

The correlation of these questions is positive and weak. This fact shows that as the importance of Destination Marketing and Management Organizations in the promotion of cultural routes increases, it is observed that the importance of integrating technologies in cultural routes tends to increase. Of course, the reverse is also true.

Conclusions-Research Limitations-Future research

Through the literature review, the need to support urban cultural tourism from NSRF programs, the Development Fund and public–private partnerships (PPPs) was initially highlighted. This support could limit seasonality and bring social and economic benefits to the local community (Triarchi & Karamanis, 2017). After all, cultural tourism focuses on the promotion of local culture and the involvement of domestic professionals, through the safeguarding of local identity (Kontis et al., 2017). One such tourism product that needs this support is cultural routes, in the context of urban cultural tourism (Hall, 2019).

In fact, the public–private partnership in cultural tourism, a practice that would strengthen initiatives in the tourism industry, in Greece is not preferred, except perhaps for some steps that have been taken lately. The literature and research have shown that PPP is detached from time-consuming bureaucratic procedures, so that they can offer benefits to local economies, faster construction and management of tourism infrastructure, a higher level of visitor satisfaction and an increase in revenue and quality of life for the local community (INSETE, 2020). For the success of a tourist PPP, it is necessary to have a balanced, properly staffed organizational structure with a clear assignment of tasks and specific responsibilities for the parties involved. Digitalization is, in addition, an asset that can give a new impetus to alternative tourism in general, increasing the quality of the experience, bringing in more digital nomads and enhancing tourism competitiveness. The need for a shift in the direction of PPPs on urban cultural tourism issues in general, which did not exist in previous years, is also confirmed by the research. The involvement of private organizations is necessary in any case for the success of cultural routes as well, while about other sources of funding, reference is made mainly to Community and national resources. This is one of the reasons why the sample proposes the autonomy of Local Government Organizations from the state, regardless of where the source of funding comes from (Heather, 2021).

Also, the review of the literature and the research demonstrated the great need for the participation of local authorities in the planning of tourism development, through DMMOs and other development agencies that will record, organize the existing tourist infrastructure and will place the guarantees of creating new ones always within the framework of the carrying capacity and with a sustainable perspective (E.E.T.A.A., 2020). Recent research shows that urban cultural tourism and cultural routes in particular are booming in Greece, with more and more initiatives at the local and national levels taking place (Zmyślony, 2011). The cultural routes seem to introduce a new model in the management, preservation, and promotion of the country's cultural reserve, through the transformation of the operation of archaeological sites, which goes from competitive to cooperative (Peeters et al., 2018; Lamnatos & Mitoula, 2020).

In addition, cultural routes have an important role in the effective and sustainable management of cultural heritage, because they shape economically and socially sustainable experiences, involve locals, and create economic and social benefits (Mitoula et al., 2016). The OECD itself highlights the contribution of cultural routes to the development of innovation, the creation of small and medium-sized enterprises, and the development of cultural tourism products and services (Diazoma, 2022).

Finally, through the theoretical and research imaging, the need to examine a sustainable digital transformation is concluded, emphasizing, however, a balance between physical presence and virtual representation (Poulaki et al., 2021). Digital cultural routes open a new chapter in the management of cultural heritage, which offers opportunities, but also poses risks. However, the improved organization of destinations, the better use of visitors' time, the easier promotion of products, and targeting the right audience are recognized (Góral, 2016). The participants in the survey believe that sustainability should be at the core of cultural routes or uniform cultural routes should be created, aiming at sustainable tourism, environmental protection, and the promotion of urban culture. In short, the sample showed a strong positive attitude toward Green Cultural Routes, whether they knew about them or not, while at the same time showing a positive attitude toward whether cultural routes can become useful tools for promoting the sustainable development of areas (Buhalis & Costa, 2005). Finally, they underlined the need to integrate technological tools in cultural routes, such as digital maps and QR codes, but also the possibilities of tours without guides, in order to achieve the goals more quickly, transparently and effectively and to make cultural routes attractive as a tourist, but also as a sustainable product.

In conclusion, the role of the DMMOs in the promotion of cultural routes, the promotion of PPP in matters of culture and tourism, and the importance of integrating technologies in these routes, through clustering or clustering techniques, are, critically speaking, the tools to which local authorities should turn. Through them, data (infrastructure, tourist facilities) can be created and utilized for each Municipality. Clusters, finally, can be used as landmarks in a cultural route, addressed to expanded groups of visitors, thus becoming a field of creativity, which would serve as a vehicle for stimulating the economic and tourist development of the region and the country.

Regarding the research limitations, although the size of the sample can be considered small, its representativeness was relatively high, because the relationship of those who answered the questions was not only direct to tourism and culture but are individuals who are in key decision-making positions. However, the results of the survey cannot be generalized. Another limitation concerns the research tool, the questionnaire. Participants may not have honestly answered the questions, a fact that often arises either from a lack of time and willingness to participate or because some of them were unable to know the answers to a very specialized and demanding questionnaire, with the prerequisite of knowledge about cultural routes and cultural tourism. The results obtained from a qualitative survey with semi-structured interviews would probably have greater credibility. Based on what has been presented so far and specifically, considering the limitations and conclusions of the research, it is proposed to design new research, which will focus on the factors that are a brake on the smooth cooperation and coordination of actions between institutions and professionals involved in the tourism and cultural sector. The aim is to evaluate both the scope of activity and the range of decision-making of those involved, the tools they use for the design of cultural products, the communication and cooperation methods they apply, and to communicate the procedures followed in case of undertaking a cultural tourism project.

References

Angelidou, M., Karachaliou, E., Angelidou, T., & Stylianidis, E. (2017). Cultural heritage in smart city environments. *International Archives of the Photogrammetry, Remote Sensing and Spatial Information Sciences – ISPRS Archives*, *42*(2W5), 27–32. https://doi.org/10.5194/isprs-archives-XLII-2-W5-27-2017

Ashworth, G., & Page, S. J. (2011). Urban tourism research: Recent progress and current paradoxes. *Tourism Management, 32*(1), 1–15. https://doi.org/10.1016/j.tourman.2010.02.002

Banila, E. (2016). *The concept of cultural route as a means of highlighting and protecting cultural heritage* (Diploma thesis, Panteion University).

Berry, D. M. (2012). *Understanding digital humanities*. https://doi.org/10.1057/9780230371934

Briedenhann, J. (2011). Economic and tourism expectations of the 2010 FIFA World Cup–A resident perspective. *Journal of Sport & Tourism, 16*(1), 5–32.

Briedenhann, J., & Wickens, E. (2004). Tourism routes as a tool for the economic development of rural areas – vibrant hope or impossible dream? *Tourism Management, 25*, 71-79.

Buhalis, D., & Costa, C. (2005). Tourism business frontiers: Consumers, products and industry. *Tourism Business Frontiers: Consumers, Products and Industry*. https://doi.org/10.4324/9780080455914

Charalabopoulou, P. (2019). *The role of cultural routes as a tourist product in the management of cultural heritage* (Diploma thesis, Harokopio University).

Charami, S. (2021). *Cultural routes: What does the plan to enhance cultural tourism include*. www.monopoli.gr/2021/05/10/istories/art-culture-sub/472978/politistikes-diadromes-ti-perilamvanei-to-sxedio-enisxysis-tou-politistikou-tourismou/

Clarke, J. (1997). Framework of approaches to sustainable tourism. *Journal of Sustainable Tourism, 5*(3), 224–233. https://doi.org/10.1080/09669589708667287

Cohen, L., Manion, L., & Morrison, K. (2008). *The methodology of educational research*. Metaichmio.

Commission of the European Communities. (2006). A renewed EU Tourism Policy: Towards a stronger partnership for European Tourism. *Com, 134*, 1–10.

Connor, S. (2004). Topologies: Michel Serres and the shapes of thought. *Anglistik, 16*, 105–107.

Creswell, J. W. (2016). *Research in education: Design, conduct and evaluation of quantitative and qualitative research*. Ion.

D'Auria, A. (2009). Urban cultural tourism: Creative approaches for heritage-based sustainable development. *International Journal of Sustainable Development, 12*(2–4), 275–289. https://doi.org/10.1504/IJSD.2009.032782

De Araujo, L. M., & Bramwell, B. (2002). Partnership and regional tourism in Brazil. *Annals of Tourism Research, 29*(4), 1138–1164. https://doi. org/10.1016/S0160-7383(02)00033-6

Diazoma. (2022). *Urban entrepreneurship study for the creation of a cluster of enterprises in the Municipality of Heraklion*. www.diazoma.gr/en/ancient-theatres-progress-works/

Duarte-Duarte, J. B., Talero-Sarmiento, L. H., & Rodríguez-Padilla, D. C. (2021). Methodological proposal for the identification of tourist routes in a particular region through clustering techniques. *Heliyon, 7*(4), e06655. https://doi. org/10.1016/j.heliyon.2021.e06655

ec.europa.eu. (2020). *Green cities fit for life*. https://ec.europa.eu/environment/european greencapital/about-the-award/faqs/

Economou, A. (2019). Digital cultural routes through geographic information systems (GIS). *Sustainable Development, Culture, Traditions, 1a*, 56–62. https://doi.org/10.26341/issn.2241-4002-2019-1a-4

E.E.T.A.A. (2020). *Tourism planning and promotion. Handbook for local authorities*. www.eetaa.gr/category/ekdoseis-eetaa/

Eichelberger, S., Peters, M., Pikkemaat, B., & Chan, C. S. (2020). Entrepreneurial ecosystems in smart cities for tourism development: From stakeholder perceptions to regional tourism policy implications. *Journal of Hospitality and Tourism Management, 45*, 319–329. https://doi. org/10.1016/j.jhtm.2020.06.011

The European Parliament. (2021). *Report on new prospects and new challenges for sustainable European tourism*. www.europarl.europa.eu/doceo/document/A-6-2005-0235_EN.html
Fergadakis, T.-I. (2021). *The importance of marketing in cultural routes as a vehicle for improving traffic, entrepreneurship and employment: The case of the Cultural-Tourist Route of Thessaly* (Diploma thesis, Hellenic Open University).
Góral, A. (2016). *Cultural routes as a medium for changes in regions*. Conference: 5th International Conference on Heritage and Sustainable Development.
Gosal, A., Geijzendorffer, I., & Václavík, T. (2019). Using social media, machine learning and natural language processing to map multiple recreational beneficiaries. *Ecosystem Services, 38*.
Guo, S., Jiang, Y., & Long, W. (2019). Urban tourism competitiveness evaluation system and its application: Comparison and analysis of regression and classification methods. *Procedia Computer Science, 162*, 429–437. https://doi.org/10.1016/j.procs.2019.12.007
Hall, C. M. (2019). Constructing sustainable tourism development: The 2030 agenda and the managerial ecology of sustainable tourism. *Journal of Sustainable Tourism, 27*(7), 1044–1060. https://doi.org/10.1080/09669582.2018.1560456
Heather, K. (2021). *Work and wander: What today's digital nomads signal for the future of travel*. www.adventuretravelnews.com/work-and-wander-what-todays-digital-nomads-signal-for-the-future-of-travel
INSETE. (2020). *Investments in public infrastructure to improve the competitiveness of the tourism sector*. www.insete.gr
Kaczorowska-Ireland, A. (2021). The Treaty of Lisbon. *European Union Law, 147*, 70–91. https://doi. org/10.4324/9780203886472–11
Kakoutis, G. (2019). *The development of cultural routes: The case of the Cultural Route of the Silk Road* (Diploma thesis, Hellenic Open University).
Kontis, A. P., Skoullos, S., & Stavrinoudis, T. (2017, October 19–21). *Marketing the local authenticity for sustainable tourism development: The case of we do local project*. Proceedings (abstracts) of the IMIC 2017 International Scientific Conference Tourism: Trends, Prospects and Implications for Enterprises and Destinations.
Koumelis, Th. (2006). *Public-private partnership for tourism promotion*. https://traveldailynews.gr/news/article/28461
Lagos, D. G. (2012). Local governance and local authorities on urban tourism development in Greece. *Sustainable Development, Culture, Traditions, 1*, 51–60.
Lamnatos, I., & Mitoula, R. (2020). City branding and sustainable development: As a case study: The historical and commercial center of Athens – 1st Municipality Community of the Municipality of Athens. *Sustainable Development, Culture, Traditions Journal, 1*(B), 46–57. https://doi. org/10.26341/issn.2241-4002-2020-1b-5
Makrigiorgou, S. (2016). *The framework for the development of cultural tourism and its connection with conferences: The case of the archaeological site of Delos and the museums of Mykonos*. University of Thessaly.
Malliara, N. (2021). *NBG: A historic opportunity for Greek Tourism the Recovery Fund*. www.insider.gr/oikonomia/184192/ete-istoriki-eykairia-gia-ton-elliniko-toyrismo-tameio-anakampsis
MCA. (2021). *11th council of Europe cultural routes forum*. culture.gov.gr.
Ministry of Tourism. (2021). *Article 4 responsibilities of the Destination Management and Promotion Agency and exceptionally establishment of a single-shareholder Destination Management and Promotion Organization*. www.opengov.gr/tourism/?p=2039
Mitoula, R., & Kaldis, P. (2020). City branding and cultural routes. *Sustainable Development, Culture, Traditions Journal*, Special Volume in Honor of Professor George I. Theodoropoulos, 118–131. https://doi.org/10.26341/issn.2241-4002-2019-sv-12

Mitoula, R., Theodoropoulou, E., Kaldis, P., & Chalioris, K. (2016). The contribution of culture to sustainable development of the Municipality of Delphi. *Sustainable Development, Culture, Traditions*, *1*, 8–29.

Nilashi, M., Ahani, A., Esfahani, M. D., Yadegaridehkordi, E., Samad, S., Ibrahim, O., & Akbari, E. (2019). Preference learning for eco-friendly hotels recommendation: A multi-criteria collaborative filtering approach. *Journal of Cleaner Production*, *215*, 767–783. https://doi.org/10.1016/j.jclepro.2019.01.012

NSRF. (2015). *Circular for the planning, implementation and monitoring of integrated territorial investments as tools for the integrated spatial development of the operational programs of the programming period 2014–2020*. Ministry of Economics, Infrastructure, Shipping & Tourism. General Secretariat for Public Investments-NSRF.

OECD. (2014). OECD Tourism and the creative economy. *OECD Studies on Tourism*. https://doi. org/10.1787/9789264207875-en

OT.gr. (2021). *NBG Survey – How tourism receipts will reach €36 billion by 2030*. www.ot.gr/2021/08/05/tourismos/ereyna-ete-pos-oi-touristikes-eisprakseis-tha-fthasoun-sta-36-dis-eyro-eos-to-2030

Pappa, M. (2018). *The development of urban tourism and city branding in the case of Athens* (Diploma thesis, University of Peloponnese).

Peeters, P., Gössling, S., Klijs, J., Milano, C., Novelli, M., Dijkmans, C., & Postma, A. (2018, October). *Research for TRAN committee-overtourism: Impact and possible policy responses*, 1–255. www.europarl.europa.eu/thinktank/en/document.html?reference=IPOL_STU(2018)629184

Poulaki, P., Kritikos, A., & Vasilakis, N. (2021). Digital technology as a means of promoting cultural tourism in Greece – case study: Acropolis Museum. *Sustainable Development, Culture, Traditions*, *1A*. doi: 10.26341/issn.2241-4002-2021-1a-1

Poulaki, P., Stavrakakis, I., Tarazonas, D., Vasilakis, N., & Valeri, M. (2021). Sustainable development and cultural heritage in Greece. In *Resilienza e sostenibilità. Risposte locali e dinamiche globali*. Giappichelli.

Region of Western Greece. (2016). *Strategic plan for an integrated territorial investment for other spatial strategies in the area of Messolonghi Lagoon – Aitoliko, Patra*. Operational Programme "Western Greece" 2014-2020. Patras.

Righini, E. (2020). *What is digital tourism? Much more than a definition*. www.doxee.com/blog/customer-experience/what-is-digital-tourism/

Somarakis, G., Lambrou, M., Panagiotopoulou, M., & Stratigea, A. (2016). Participatory management of cultural resources as a pillar of local development: The case of the residential unit of Kissamos. *Texts of Regional Science*, *7*(1), 129–147.

Šť'Astná, M., Vaishar, A., Brychta, J., Tuzová, K., Zloch, J., & Stodolová, V. (2020). Cultural tourism as a driver of rural development. Case study: Southern Moravia. *Sustainability (Switzerland)*, *12*(21), 1–16. https://doi.org/10.3390/su12219064

TornosNews.gr. (2021). *Konstantinos Michalos: New actions are needed, through the NSRF, in alternative tourism*. www.tornosnews.gr/stiles/prosopa/51437-chreiazontai-nees-draseis-meso-toy-espa-ston-enallaktiko-toyrismo.html

Triarchi, E., & Karamanis, K. (2017). Alternative tourism development: A theoretical background. *World Journal of Business and Management*, *3*(1), 35. https://doi.org/10.5296/wjbm.v3i1.11198

Zmyślony, P. (2011). Globalization, tourism and cities: Pros and cons. *Folia Turistica*, *25*(1), 299–312.

INDEX

Note: Page numbers in *italics* indicate a figure and page numbers in **bold** indicate a table on the corresponding page.